Electric Power Conversion and Micro-Grids

Edited by Majid Nayeripour and Mahdi Mansouri

Published in London, United Kingdom

IntechOpen

Supporting open minds since 2005

Electric Power Conversion and Micro-Grids
http://dx.doi.org/10.5772/intechopen.91563
Edited by Majid Nayeripour and Mahdi Mansouri

Assistant to the Editors: Mohammad Ali Ghaderi

Contributors
Hadis Hajebrahimi, Tapparit Bangtit, Reham M. Abdel Fattah, Mohamed A. Ebrahim, Ebtisam M. Saied, Samir M. Abdel Maksoud, Hisham El Khashab, Emmanuel Hernandez Mayoral, Carlos D. Aguilar Gómez, Efraín Dueñas Reyes, Reynaldo Iracheta Cortez, Omar Rodríguez Rivera, Wilder Durante Gómez, Edwin F. Mendoza, Christian R. Jiménez Román, José I. Barreto Muñoz, Carlos J. Martínez Hernández, Juan D. Rodriguez Romero, Borys Pleskach, Sekhane Hocine, Sajjad Makhdoomi Kaviri, Suzan Eren, Alireza Bakhshai

First published in London, United Kingdom, 2022 by IntechOpen
IntechOpen is the global imprint of INTECHOPEN LIMITED, registered in England and Wales, registration number: 11086078, 5 Princes Gate Court, London, SW7 2QJ, United Kingdom
Printed in Croatia

British Library Cataloguing-in-Publication Data
A catalogue record for this book is available from the British Library

Additional hard and PDF copies can be obtained from orders@intechopen.com

Electric Power Conversion and Micro-Grids
Edited by Majid Nayeripour and Mahdi Mansouri
p. cm.
Print ISBN 978-1-83969-388-5
Online ISBN 978-1-83969-389-2
eBook (PDF) ISBN 978-1-83969-390-8

For EU product safety concerns:
IN TECH d.o.o., Prolaz Marije Krucifikse Kozulić 3, 51000 Rijeka, Croatia, info@intechopen.com or visit our website at intechopen.com.

Meet the editors

After 8 years of industrial experience and academic work in the electrical engineering and renewable energy fields, Prof. Majid Nayeripour was promoted to full professor in the field of micro-grids in 2016. He was given a sabbatical from Shiraz University of Technology, Iran, and was invited to Cologne University of Applied Sciences, Germany in January 2016. During his research, he gained new experiences about problems relating to high penetration levels of distributed generations and toward having 100% renewable energy in Germany, and as a result, he was awarded a fellowship program for an experienced researcher from the Alexander von Humboldt (AvH) Foundation in 2017. Currently, he is at the Cologne University of Applied Science and is involved in research on the control and dynamic investigation of interconnected micro-grids. He has published more than 120 journals and conference papers, five books, and supervised more than ten research projects.

Dr. Mahdi Mansouri was born in Yazd, Iran in 1975. He received both his B.S. degree in electronic engineering and MSc degree in electronic power from the Sharif University of Technology at Power Electronics-STATCOM and his Ph.D. degree in renewable energy systems from doubly fed induction generator (DFIG)-based wind turbines from the Shiraz University of Technology. He has 20 years of experience in high-voltage transmission substations and lines as a technical engineer, a consultant, and an executive project manager. His research interests include flexible alternating current transmission system devices, power quality, and power system protection. He currently conducts power electronics, power-relay protection, and power-quality projects as a consultant and project manager.

Contents

Preface

This edited volume is a collection of reviewed and relevant research chapters concerning developments within micro-grids and electric power conversion field of study. The book includes scholarly contributions by various authors and is edited by a group of experts in such fields. Each contribution comes as a separate chapter complete in itself but is directly related to the book's topics and objectives.

The book contains the following sections: *Micro-Grids: A Cost-Efficient Controller for Isolated Ones in Remote Areas*, *A New Energy Management Method for Energy Storage Systems in Micro-Grids*, *Power Quality in Micro-Grids with Energy Storage Technologies*, *A Study on Micro-Grids toward 100% Renewable Energy*, *Salp Swarm Optimization,* and *Estimation of Hidden Energy Losses in Technological Systems*. The target audience comprises scholars and specialists in the field.

Majid Nayeripour
Cologne University of Applied Sciences,
Germany

Alexander von Humboldt (AvH) Foundation,
Bonn, Germany

Mahdi Mansouri
Yazd University,
Iran

Section 1

Micro-Grids: A Cost-Efficient Controller for Isolated Ones in Remote Areas

Chapter 1

Design and Simulation of Low-Cost Microgrid Controller in Off-Grid Remote Areas

Tapparit Bangtit

Abstract

This study presents the microgrid controller with an energy management strategy for an off-grid microgrid, consisting of an energy storage system (ESS), photovoltaic system (PV), micro-hydro, and diesel generator. The aim is to investigate the improved electrical distribution and off-grid operation in remote areas. The off-grid microgrid model and the control algorithms developed using MATLAB Simulink and State flow. The energy management system is focusing on the state of charge of the energy storage system. The microgrid controller controls the operation mode and power generation from the distributed generations' local controller, i.e., PV, micro-hydro, and diesel. It also controls the smart meters of the loads to be connected or disconnected to the microgrid. The simulation results show that the proposed microgrid control can control the target off-grid microgrid in given possible scenarios. The off-grid microgrid managed to meet the energy demand with the lowest power outage and the diesel generator operation's lowest cost.

Keywords: Microgrid, Microgrid Control, Microgrid Controller, Off-grid Microgrid, Islanded Microgrid, Remote Microgrid. Low-cost microgrid controller. Renewable energy, Energy Management

1. Introduction

In remote areas of some countries worldwide, an approaching the electricity is sometimes unable for the communities located away from the central utility system. In Thailand, most of the villages without electricity located in remote jungle areas, which difficult to access and on the sea island. Therefore, expanding the service areas by planting power poles and cables is challenging because several problems and procedures for permitting the areas could take a long time. Additionally, The Ministry of Interior supports the monetary fund for distributing the Solar Home System (SHS) to around two hundred thousand households in Thailand's rural areas, which inaccessible to the primary utility grid [1]. Integrating the local electricity generation, i.e., micro-hydro plants and photovoltaic systems with battery energy storage controlled with a microgrid controller, might be the practical and optimal solution. The power generation and consumption in the off-grid microgrid could be managed intelligently by the microgrid controller. The proper microgrid controller algorithm could improve the efficiency and reliability of the microgrid. Besides, the proposed microgrid controller could apply in other areas, such as islands and the area between Thailand and Myanmar's burden. Therefore, this project aims to study and develop microgrid control algorithms for a low-cost off-grid microgrid controller. The controller would have required low maintenance and improved its operation conveniently to be implemented in PEA projects in Thailand and reduce importing the microgrid controller from other countries.

2. Literature review

2.1 Microgrids

A microgrid is a subsystem of the primary electrical grid, which generally comprises generation capabilities, storage devices or energy storage systems (ESS), and controllable loads. Additionally, a microgrid may operate either connected to the primary electrical grid, grid-connected/on-grid mode, or operated independently from the primary electrical grid, islanded/off-grid mode [2, 3]. The U.S. Department of Energy also defines the microgrid as "a group of interconnected loads and distributed energy resources (DERs) with clearly defined electrical boundaries that acts as a single controllable entity for the grid and can connect and disconnect from the grid to enable it to operate in both grid-connected or islanded modes" [4]. The basic concept of a microgrid is illustrated in **Figure 1**.

Microgrids have crucial advantages, including (1) There is no disruption to the loads within the microgrid systems due to abilities to separate itself when disturbances occur in the primary grid. (2) The optimization of the system operation is contributed. (3) Grid failure during peak load periods is prevented by reducing the load on the grid. (4) Zero or low emission electricity generation provides benefits to the environment [5].

2.2 Off-grid microgrid

When the electricity from the primary utility grid is inaccessible and costly to expand the system to remote areas, off-grid microgrids or standalone microgrids are commonly be the solution [6].

In recent years, research on the off-grid or standalone microgrid has become interesting for researchers and electric utilities worldwide. Many studies have shown that an islanded microgrid operation is feasible and has many economic and

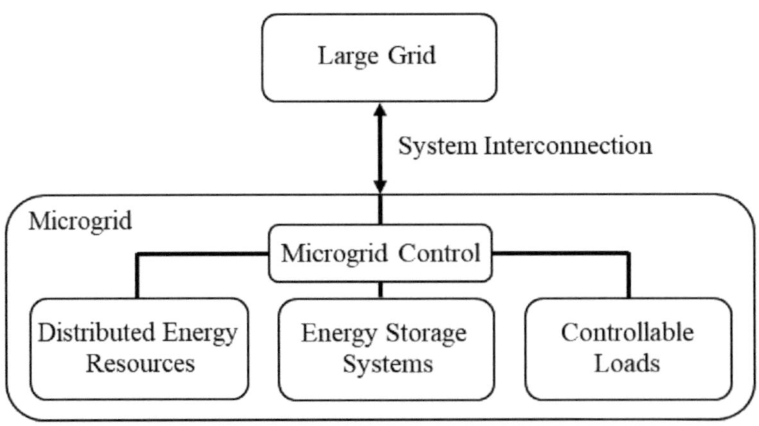

Figure 1.
Basic concept of a microgrid (adapted from [2]).

technical benefits of the pilot project in different countries in the world [7]. In south Asia, solar photovoltaic (PV) and mini/micro hydro systems technologies are usually used for off-grid electrification. Solar Home Systems (SHSs) and mini-grid are typically included 20–100 Wp (peak watt) PV panels, batteries for energy storage, and high-efficiency lamps, e.g., LED. The systems typically supply AC electricity in the range of 2–150 kWp, mainly used for lighting. The mini/micro hydro systems (typically range 50 kW – 3 MW) also have been used to form mini/microgrids to provide AC electricity locally. The rural electrification in detail and renewable energy resources in South Asia, particularly in India, Bangladesh, and Nepal, is mentioned in [6].

In [8], the current development state of major off-grid microgrids worldwide has been presented. In China, 50% of the population are living in rural areas. By 2013, the capacity of grid-connected and off-grid microgrids comprised of solar PV, hydro, and diesel was increased. China also aims to install PV off-grid microgrids to provide 1.19 million people who without electricity in 2015. In India, one-quarter of India's population still without electricity. EV batteries are promoted to be used as a grid source by the Indian government. They have biomass gasifiers, solar PV, small wind turbines, small hydro plants, and biogas systems to supply the electricity in rural off-grid and industrial applications. In the Philippines, there are 375 MW in the total capacity of diesel off-grid microgrids that have less than 500 kW operating capacity to supply electricity to rural areas. These microgrids are running around 6–8 hours per day. In Africa, two-third of the population still unable to access electricity. Solar home systems (SHSs) Off-grid microgrids are rapidly being developed, especially in Sub-Saharan. On-grid microgrids in Sal Island and Santiago have a mix of various generations, including solar PV, wind turbine, and diesel. In this country, PV and diesel generators are being funded to scaling up continuously [8]. **Figure 2** illustrates an example of the off-grid microgrid system.

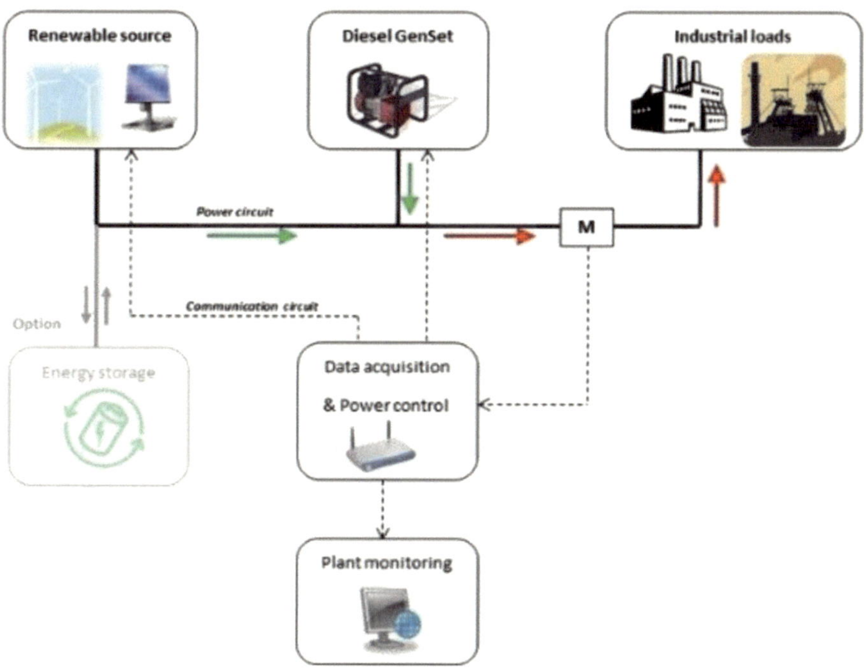

Figure 2.
An example of off-grid microgrid system [9].

2.3 Distributed generations

The definition of the distributed generation has a wide range of generation schemes. Some countries focus on large-scale generation connected to the grid. On the other hand, some other countries focus on small-scale generation units connected to the grid. According to [10], distributed generation has been defined as 'an electric power generation source connected directly to the distribution network of on the customer side of the meter'. Implementing distributed generations in the distribution system has many benefits, including:

1. Support load increase locally.
2. It can be installed easily.
3. Flexible for location and voltage level.
4. Well sized for small required load.
5. Economical for remote or standalone areas.
6. Reduce wholesale price due to decrease in main grid demand.
7. Increase the systems' equipment lifetime.
8. Reduce losses in transmission.
9. Help load management and peak shaving.
10. Increase or maintain system stability.
11. Can be the stand-by supply in case of emergency.
12. Release more transmission capacity in transmission lines.
13. Eliminate or reduce the emission to the environment [11].

The example of distributed generation technologies including reciprocating engines, gas turbines, microturbines, small/micro-hydro, fuel cells, photovoltaics, solar thermal, wind turbine, geothermal, ocean energy, Stirling engine, and battery storage [10, 12]. The distributed generation technologies are illustrated in **Figure 3**.

2.3.1 Photovoltaic

Photovoltaics (PVs) are commonly known as solar panels. They comprise the multiple semi-conductor cells connected either parallel or series to generate direct current (DC) electricity using the sunlight radiation through the photovoltaic effect. Direct current electricity is converted to alternating current (AC) by typically using inverters in order to supply electricity to the AC loads or to connect to the utility grid [14, 15]. The output power is proportional to the surface area of the cells. The output current and voltage are functions of solar radiation and temperature, respectively. A maximum power point tracking (MPPT) system inside the inverter has a crucial role in obtaining the maximum power output [14]. **Figure 4**

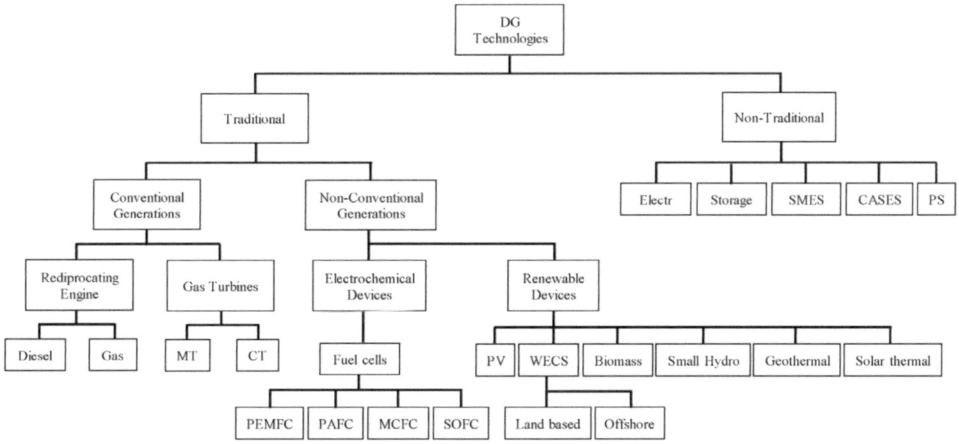

Figure 3.
Distributed generation technologies (adapted from [13]).

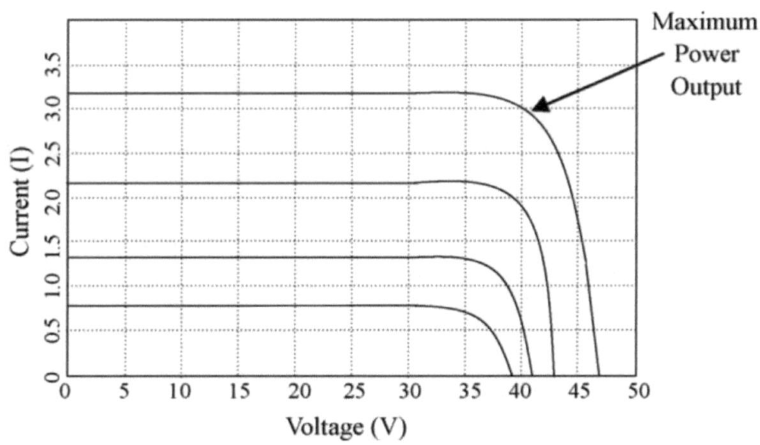

Figure 4.
Typical V-I characteristic of a PV module [14].

illustrates the typical V-I characteristics of the PV module. However, there are two significant challenges for solar PV to be connected to the electricity grid. (1). Solar PV systems lack inertia affecting the grid stability due to the lack of rotating machines, unlike other conventional rotating machines [16]. (2). Power generated from solar PVs depends on the availability and intensity of sunlight and temperature. Grid instability could occur when under or over-generation [17]. Solar PV connected to the AC grid schematic diagram is shown in **Figure 5**.

2.3.2 Micro hydro

A micro-hydropower system is considered one of the most popular in developing countries. Most of them operate in isolated mode to provide electricity to the small remote or rural areas where the primary utility grid cannot reach or not feasible due to restriction and financial-economic issues [19]. Micro-hydro power plant (MHPPs) technology has recently proven to be a feasible electric generation with good performance and low investment cost [20].

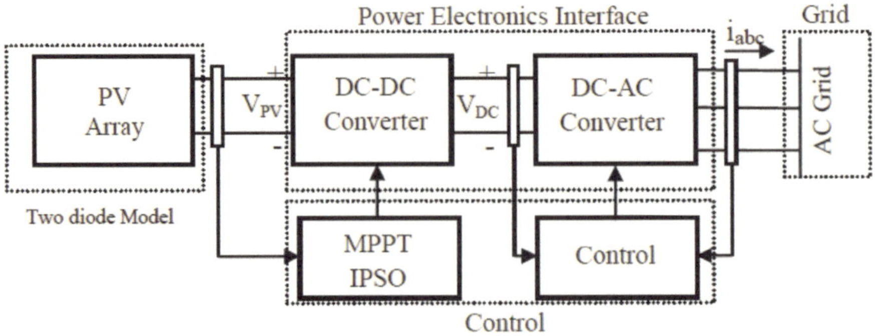

Figure 5.
Block diagram of the PV system connect to the AC grid [18].

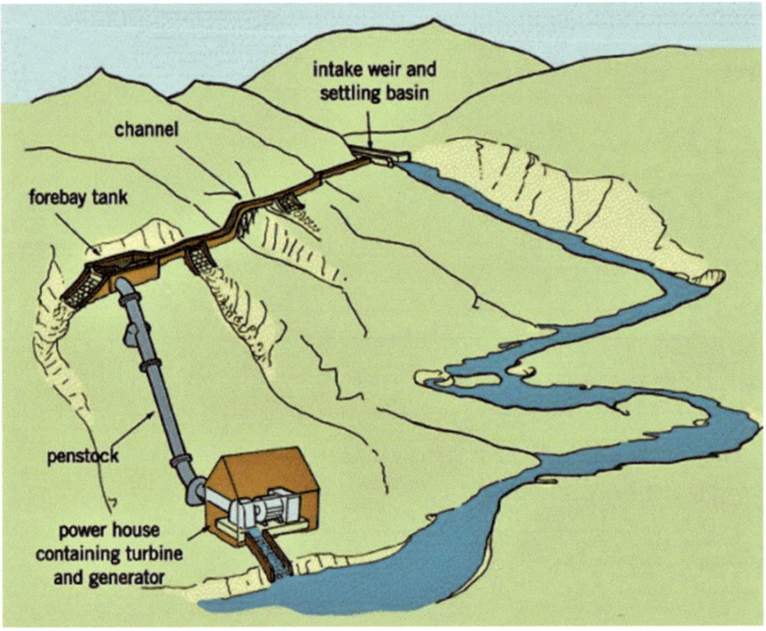

Figure 6.
Small hydro scheme [21].

Small hydropower plant can be divided by the power capacity:

1. Mini hydro: less than 1,000 kW capacity.

2. Micro-hydro: less than 100 kW capacity [19].

A typically small hydro scheme for medium or high head is shown in **Figure 6**. Water from the river is taken from a weir through an intake then passes through a settling tank or forebay to suspend particles. Then, a penstock, a pressure pipe, conveys the water to the turbine to generate the electricity and then discharges down back to the river.

Hydro-turbines convert water pressure or water speed into mechanical power in the turbine shaft, connected to the generator. The generator converts mechanical power to electrical power. The general output power can be expressed as:

$$P = \eta \rho g Q H \tag{1}$$

Where, P is the mechanical power produced at the turbine shaft (Watts), η is the hydraulic efficiency of the turbine, ρ is the density of water (kg/m^3). G is the gravity acceleration (m/s^2), Q is the volume flow rate passing through the turbine (m^3/s), and H is the effective pressure head of water across the turbine (m) [21].

In [20], A design and simulation of the micro hydropower plant (MHPP) connected to the grid with three-phase power electronics control with maximum power point tracking (MPPT) is presented. In [22], The variable speed operation technique of micro-hydropower generation system (MHPGS) using a direct drive Permanent Magnet Synchronous Generator (PMSG) topology has been verified effectiveness by obtaining more power from the hydraulic turbine. The study is using the three-phase grid-connected converter, which can be controlled in grid-feeding mode. This topology can be used in the off-grid microgrid because the required power from the microgrid controller can be set to the power reference in the power management at the grid-side converter. The control architecture of a variable speed MHPGS is shown in **Figure 7**.

2.3.3 Diesel generator

Diesel generators in the hybrid PV diesel systems have been explained in [23]. Diesel Generators and combustion engines are usually used as the backup electrical energy supply when the primary electrical utility grid cannot provide electricity. It is usually used to be the primary source to maintain the standalone system's voltage and frequency. Diesel generators or diesel generator sets can operate individually or connected with other energy resources such as solar PV. A dual diesel generator system including smaller and larger generators is employed for fuel-saving. The small generator operates when light loads. In contrast, the larger operates when the loads increased. Diesel generators have low efficiency at low load due to the non-linear to load ratio of the fuel consumption. The maintenance period is based on the operational hours. Additionally, electricity generation from the diesel generator sets needs expensive fuel prices, and there are also maintenance and operating costs. In the case of a hybrid system with high penetration of PV, energy storage is needed, and diesel generator sets can be operated in either continuous or intermittent operation.

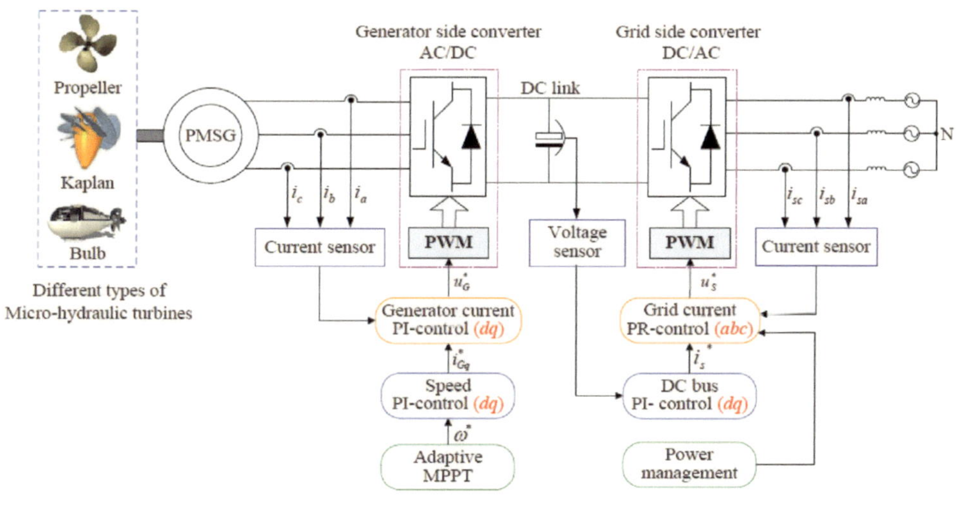

Figure 7.
Architecture of a variable speed MHPGS and its control design [22].

1. Continuous operation: Diesel generator set must be maintained to not operate below the minimum generation. The control of PV penetration and dump loads is needed to maintain the energy generated from the diesel generator sets.

2. Intermittent operation: Diesel Genset can be controlled to be ON or OFF by the controller, which is normally based on PLC [23, 24].

A control schematic of the synchronous diesel generators is shown in **Figure 8**. The mechanical power and field current is controlled to regulate the voltage amplitude and frequency, respectively [25].

2.4 Energy storage system

Energy Storage System (ESS) plays an essential role in the microgrids. It has an essential role in balancing the power and energy demand from the loads with power and energy generation from the distributed resources such as solar PV and wind turbine, which rely on weather conditions. Thus, energy storage can allow energy to be stored during high renewable generation or low demand periods and used during low renewable production or high demand periods [26]. The excess generated energy can be stored in the ESS and then supply to the loads during high demand or when the load demand fluctuates. Furthermore, ESS supports intermittent DGs dynamic and seamlessly transitions the microgrid between grid-connected mode and islanded mode. However, expensive devices requiring maintenance and more space for battery banks are the major drawbacks of the ESS [5].

Energy storage is essential for power and voltage smoothing, energy management, frequency regulation, peak shaving, load leveling, seasonal storage, and standby generation during a fault [27]. Energy storage technologies can be summarized as follows:

1. Pumped Hydro Storage (PHS).

2. Compressed Air Energy Storage (CAES).

3. Batteries.

4. Hydrogen Energy Storage (Fuel Cells).

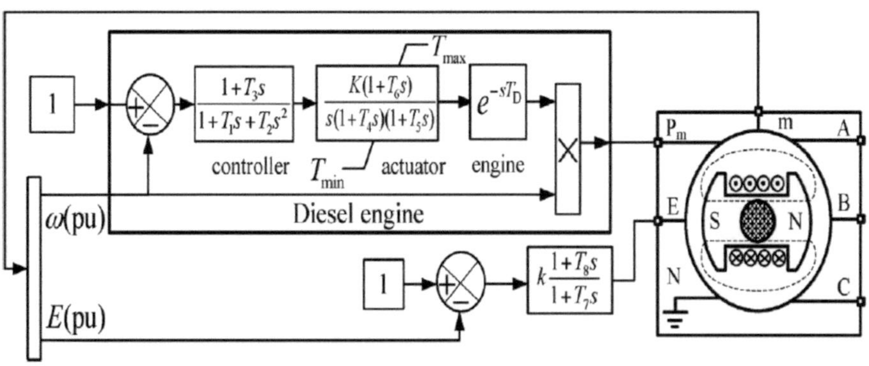

Figure 8.
Control of the diesel engine and synchronous generator combination [25].

5. Thermal Energy Storage (TES).

6. Superconducting Magnetic Energy Storage (SMES).

7. Flywheel Energy Storage (FES).

8. Supercapacitor/Ultracapacitor [28].

2.5 Loads

Typically, the population resides in the remote, isolated areas are generally small-sized communities, for instance, forests-villages and islands. These communities typically have low income, which unattractive and costly to expend the electricity from the primary utility system to reach those areas [6]. Loads in the microgrid are typically categorized into two types:

1. Fixed loads: In normal operating conditions, fixed loads cannot alter and must be satisfied,

2. Flexible loads: When the economic incentives or islanding requirement, the flexible load or adjustable or responsive loads could be shed or curtailed and deferred to meet the requirement [4].

2.5.1 Load shedding

Generally, for an islanded or off-grid microgrid with high renewable energy penetration, load shedding is recognized as a necessary way to maintain the power balance and stability in case of emergency conditions or if the total power generation is not enough to meet the total demand [29]. Many studies have focused on finding a suitable load shedding scheme. In [30], two types of load shedding were formed: (1) centralized load shedding and (2) the distributed load shedding.

2.6 Microgrid control

2.6.1 Hierarchical control

Microgrids that combine many different distributed generation and energy storage devices should have capabilities to import/export power to/from the grid and support grid-connected and standalone applications.

Hierarchical Control can be divided into three categories:

1. Primary Control (droop control).

   Primary Control has a role in assuring power sharing among the distributed generations. The frequency and voltage magnitude are adjusted. The droop control in each external inverter allows DGs to operate autonomously based on the local measurement. The communication in this control scheme is avoided.

2. Secondary Control (V, F control).

   Secondary Control has tasks to restore the frequency and voltage deviation produced by virtual inertia and virtual impedances. Also, perform synchronization between the microgrid and the primary grid.

3. Tertiary Control (P, Q control).

   Tertiary Control adjusts the microgrid inverter's set point to control the power flow between the microgrid and the primary grid [5]. The hierarchical architecture of a microgrid is illustrated in **Figure 9**.

2.6.2 Outer control loops

The power converter can be connected to the grid in parallel, and the power between the DGs and the grid can be controlled to meet the desired power. The power converters can be controlled separated into three classes:

1. Grid-forming: The converter acts as an ideal AC voltage source balancing the power generation and loads with fixed frequency and voltage amplitude. It is designed for autonomous operation, which can provide a reference for the voltage and frequency.

2. Grid-feeding: The converter acts as an ideal current source, which its voltage and frequency follow the connected grid. It is designed to deliver a specific amount of active and reactive power to an energized grid. It is actually used with mostly DGs in the microgrid, such as solar PV and wind turbine.

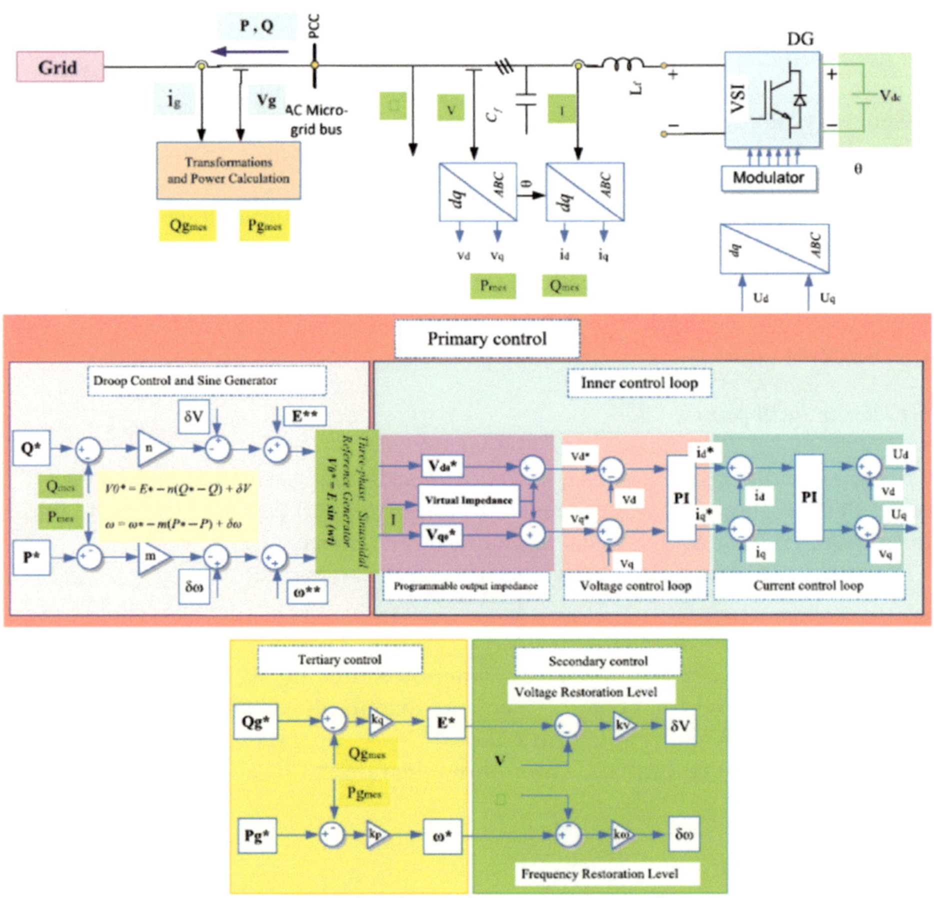

Figure 9.
Hierarchical architecture of a microgrid [5].

Nonetheless, grid-feeding cannot operate independently without an energized grid.

3. Grid-supporting: The converter the voltage amplitude with reactive power and control frequency with active power by emulating the artificial droop control equivalent to the droop control in synchronous generators the in utility grid. It is designed to support the regulation and stability of the microgrid. Grid supporting can operate as a voltage source and current source [5, 31]. **Figure 10** illustrates the converter categories. The classification of grid-connected is presented in **Table 1**.

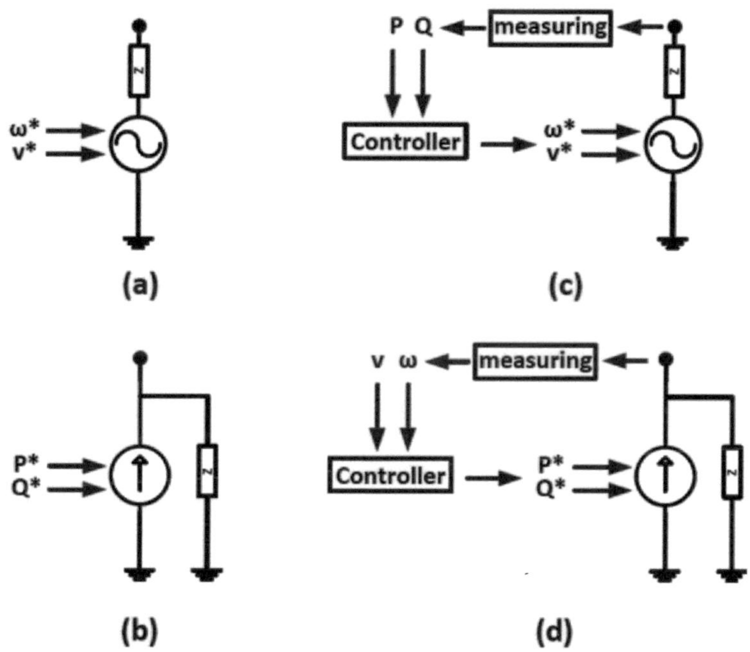

Figure 10.
Converter categories; (a) grid-forming; (b) grid-feeding; (c) grid-supporting voltage source; (d) grid-supporting current source (Adapted from [27]).

Contribution to the grid	Grid-forming	Grid-feeding	Grid-supporting
Source type	Ideal voltage source	Ideal current source	Non-ideal voltage or current source
Control type	Constant frequency/ voltage control	PQ control	Droop control
Combination	Series	Parallel	Parallel or series
Output impedance	$Z_d = 0$	$Z_d = \infty$	Finite, nonzero
Output frequency	Fixed frequency	Grid synchronized	Frequency droop
Application	Isolated	Grid-connected	Grid-connected or Isolated

Table 1.
The classification of grid-connected according to their electrical behavior and their contribution to the grid [5].

2.6.3 Inner control loops

Different control techniques can be used for the inner control loops for distributed generations to control the voltage source inverter, including (1). Classical PID, (2). Proportional resonant (PR), (3) Predictive control, (4). Dead-beat control, (5). Hysteresis control, (6). LQG/LQR, (7) Sliding mode controls, (8). H∞ controller, (9) Repetitive controller, and (10) Neural networks and fuzzy control methods [5].

2.6.4 Microgrid controller

2.6.4.1 Dynamic dispatch strategy

According to [32, 33], the control strategy of the hybrid system is divided into two categories:

1. Dynamic strategy: to maintain the system stability by focused on voltage magnitude and frequency of the system. The Timestep is less than a second.
2. Dispatch strategy: to maintain a power balance in the system based on power flow measurements, follow the algorithm which facilitates the interaction among the distributed generations, energy management devices, and loads. Time step ranges from minutes to hours.

2.6.4.2 Existing dispatch strategies

The existing power management strategies reviewed in [33] can be summarized as follows:

- Peak Shaving: Battery is used as a buffer for the load demand fluctuation. It can be charged by the excess power from the distributed generations but will be charged by a diesel generator when the diesel generator's minimum load condition is reached.
- Load Following: The load demand is met by the distributed generation's power, including a diesel generator. The battery will be charged only when the minimum load condition of the diesel generator is reached.
- Frugal discharge: When the load demand is more than the total power generation from the distributed generations, the load demand will be supplied by the battery. A diesel generator will operate to support load demand only when more cost-effective than discharged the battery.
- SOC set-point: Diesel generator always operates at full power for maximum efficiency. The excess power charges the battery from the diesel generator up to the defined SOC.
- Full power/minimum run time: to prevent the frequently start cycles of the diesel generator. The diesel generator is operated at full power. The battery is mainly used for storing excess power from other distributed generations.

2.6.4.3 Intelligent-based power management

Intelligent-based power management uses computational intelligence techniques that utilize technology and computer science to mimic nature and human beings. The example of computational intelligence are:

- Fuzzy logic control (FLC).
- Artificial neural networks (ANN).
- Adaptive neuro-fuzzy inference system (ANFIS).
- Genetic Algorithms (GA).
- Swarm-based optimisation methods.

In [34], it is concluded that there is no universal optimisation algorithm to be the best in solving optimisation problems. Hybrid optimisation techniques can utilize the strong advantages of each optimisation algorithm to reach an optimal solution.

3. Methodology

In this study, software MATLAB R2020a with Simulink toolbox is used to develop the microgrid system model, including energy storage system, distributed generations, loads. The MATLAB Stateflow toolbox is used to develop the algorithm for the microgrid controller. All of the simulations in this study are conducted in the Phasor mode with the frequency at 50 Hz., Solver: ode23tb (stiff/TR-BDF2) with type: variable-step.

3.1 System layout

In this study, The off-grid microgrid model using this study consists of the distributed generations (DGs), i.e., photovoltaics (PVs), micro-hydro, diesel

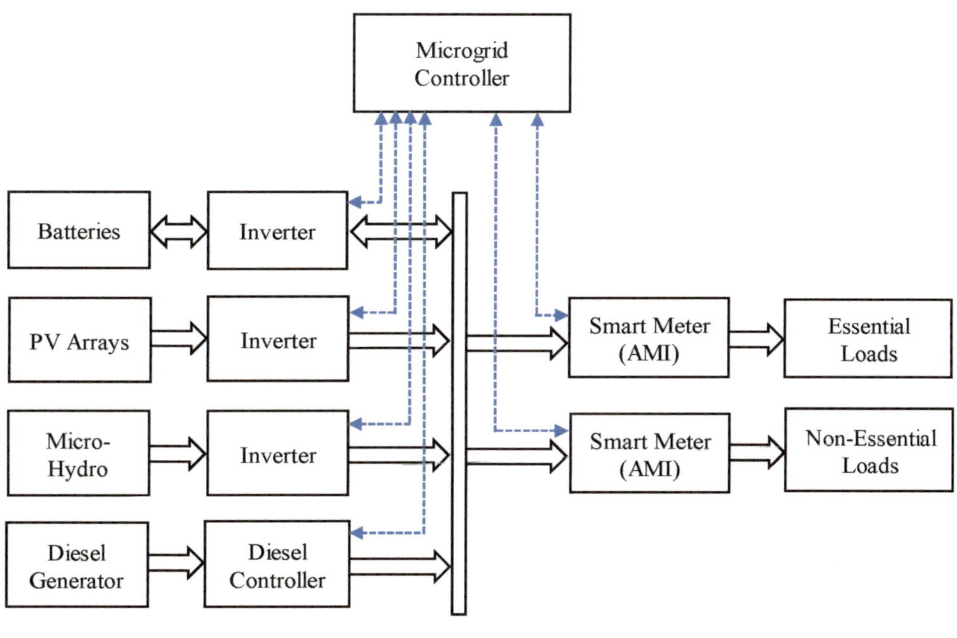

Figure 11.
Block diagram of the off-grid microgrid.

Genset. The loads can be divided into two groups; (1) Essential-loads, i.e., a hospital, a school, and agriculture pumps. (2) Non-essential loads, i.e., houses (residential loads). The off-grid microgrid has an energy storage system (ESS) connected to the system.

Figure 11 shows the block diagram of off-grid microgrid with microgrid controller, which consists of (1) energy storage system, which is batteries connected to the inverter. (2) Various distributed generation, i.e., photovoltaic arrays (PV), micro-hydro, diesel generator, has inverter as a local controller. (3) Essential and Non-Essential Loads connected to the smart meters. All elements are connected to the AC bus. The inverters, local controllers, and smart meters are monitored and controlled by the microgrid controller or microgrid central controller (MGCC) through the communication channel.

3.2 Model of simulation

3.2.1 Off-grid microgrid model

The off-grid microgrid system using in this study is shown in **Figure 12**. The energy storage system (ESS), photovoltaic (PV), micro-hydro, and the diesel generator are connected in three-phase at the 400 V bus. The V-I measurement tools measure the voltages and the currents of each distributed generation and ESS. Three-phase transformer 0.4/22 kV is connected to step up the system voltage to the medium voltage system (22 kV). The voltage and current of the bus 22 kV are also measured. The transformer, 22/0.4 kV, tr1, and tr2, are connected to step-down the voltage from 22 kV to 400Vline-line (230Vphase-neutral) to supply the loads single-phase 230Vphase-neutral. The unbalance voltage and the loss in the transmission line are neglected in this study.

3.2.2 System operation

The off-grid microgrid needs at least one voltage source to be operated in the microgrid. In this study, the energy storage system (ESS) has a responsibility as a

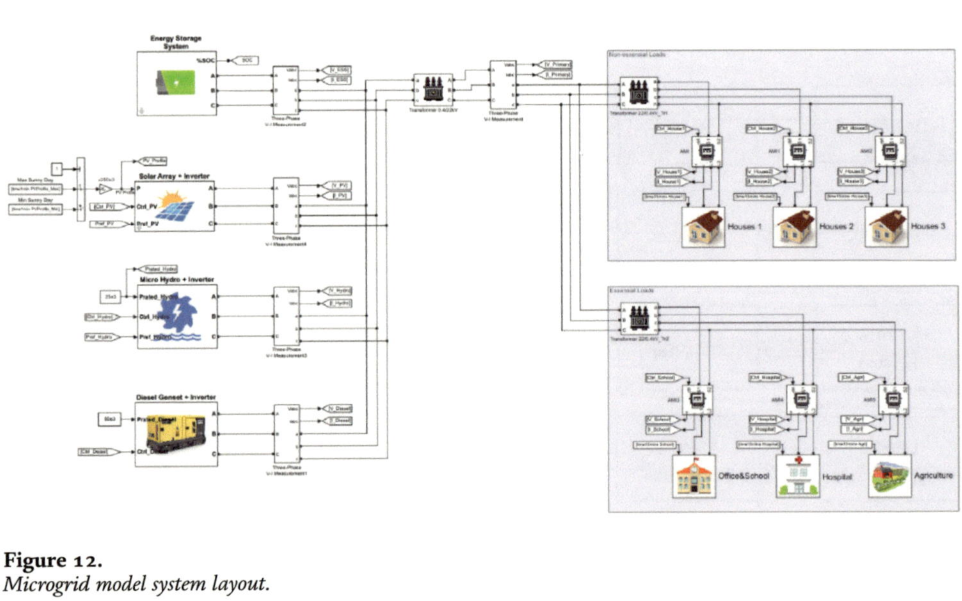

Figure 12.
Microgrid model system layout.

central component of the off-grid microgrid. The ESS always operates in the voltage and frequency (v, f) mode to be the main components that generate the reference frequency of the microgrid and maintain the microgrid's voltage and frequency. The distributed generation, i.e., PV, micro-hydro, and diesel generator, is connected to the microgrid. The inverter controlled the operating mode and power reference from the microgrid controller as a balanced three-phase supply. The microgrid central controller (MGCC) controls the local controller of each DG and loads by the proposed algorithms to meet the load demand and maintain the microgrid's voltage and frequency.

3.3 PV model

Power generated from photovoltaic arrays (PVs) is dependent on the light intensity and temperature of the PV surface. In this study, the PV model uses the Simple Inverter block in MATLAB Simulink to imitate that the PV arrays are connected to the 3-phase inverter, which can be switched the operation mode between Maximum power point tracking (MPPT) and power reference mode. The inverter is a local controller which acts as a slave to the microgrid controller. In **Figure 13**, the power generated from PV depends on the power profile (P), which can be selected to be maximum or minimum sunlight day. The microgrid controller can control the inverter's operation mode by the control signal from the microgrid controller (Ctrl_PV). When the inverter is controlled in the power reference mode, the power generated from the PV to the microgrid can be controlled by the reference power value sending from the microgrid controller (Pref_PV).

3.3.1 PV profile

In this study, the power generated from the PV model to the grid depends on the daily power profile. **Figure 14** shows the four selected power profiles from daily 365 days of power generated from real PV power generated profiles in Thailand, i.e., minimum, maximum, clear, and cloudy. The maximum is the profile from the day that the PV generated the highest power. The minimum is the profile from the day that the PV generated lowest power. The clear day is the profile from the average clear day. The cloudy day is the profile from the average cloudy day.

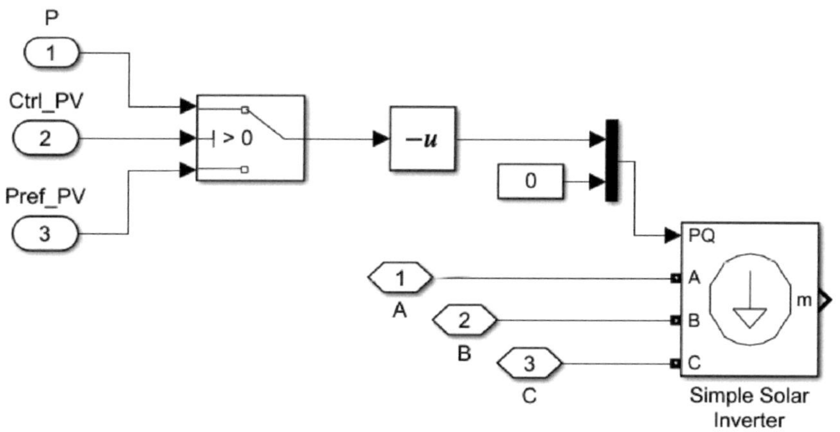

Figure 13.
PV with inverter model.

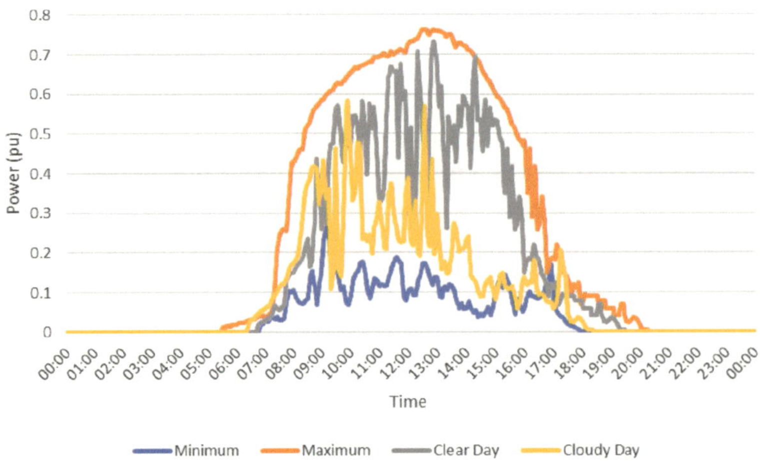

Figure 14.
Daily power profile of PV.

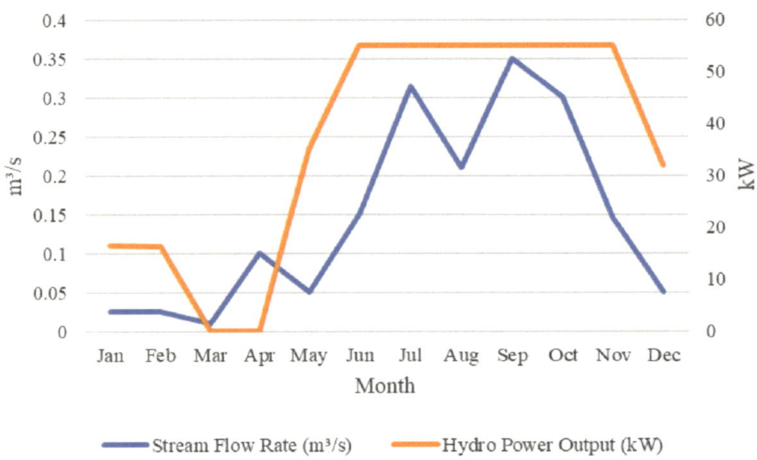

Figure 15.
Stream flow rate and power output of the micro-hydro (adapted from [35]).

3.4 Micro-hydro model

In this study, the hydro plant uses the real profile of power generation from the hydropower plant in Khunpae village, Chiangmai, Thailand. The hydropower plant has a maximum power of 55.57 kW and 36.35 kW on average [35]. The output power of the hydropower is less than 100 kW. Thus, the hydro plant in this study can be categorized to be a 'micro-hydro. The power generated from the micro-hydro depends on the water level in the weir. The micro-hydro generates maximum power output during the rainy season (Jun-Nov). In the dry season, the micro-hydro cannot generate power due to a water shortage, as shown in **Figure 15**.

The micro-hydro's local control system in this study is assumed that the micro-hydro is connected to the microgrid with a 3-phase inverter. The power injected into the microgrid is controlled by the microgrid controller. The control architecture of the micro-hydro is shown in **Figure 16**.

In this study, the micro-hydro model uses the Simple Inverter block in MATLAB Simulink to imitate the 3-phase inverter, which can be switched the operation mode between Maximum power point tracking (MPPT) and power reference mode.

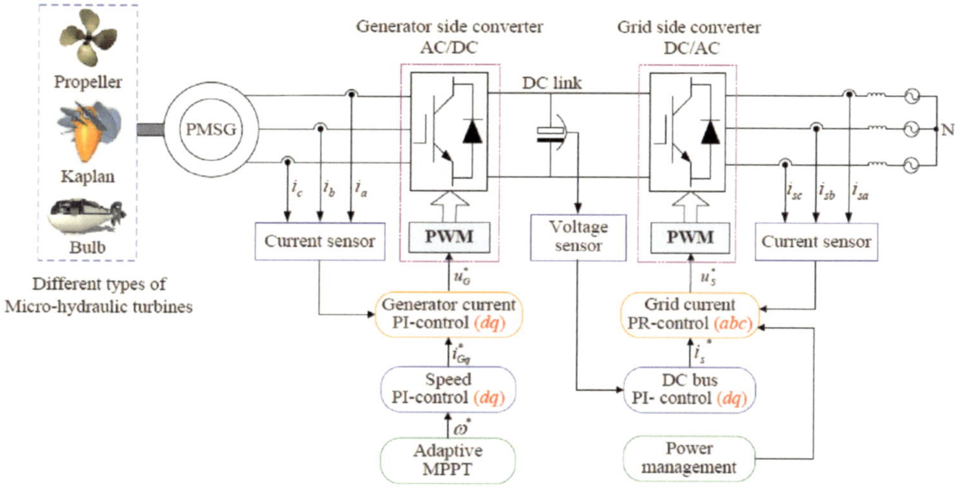

Figure 16.
Control architecture in the micro-hydro [22].

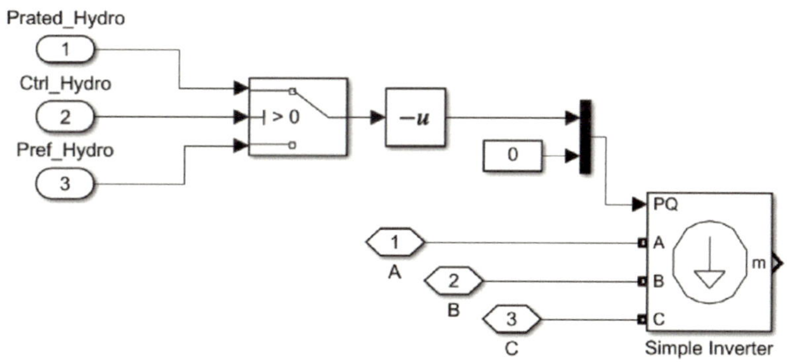

Figure 17.
Micro-hydro with inverter model.

The inverter is a local controller which acts as a slave to the microgrid controller. In **Figure 17**, the power generated from the micro-hydro is set to the rated power (Prated_Hydro). The microgrid controller can control the operation mode of the inverter of the micro-hydro by the control signal from the microgrid controller (Ctrl_Hydro). When the inverter is controlled in the power reference mode, the power generated from the micro-hydro to the microgrid can be controlled by the reference power value sending from the microgrid controller (Pref_Hydro).

3.5 Diesel Genset model

The diesel generator in this study is used to be a backup system of the off-grid microgrid. In a normal situation, the diesel generator should not be operated due to the cost of fuel. The diesel generator is modeled in MATLAB Simulink assumed that the diesel generator is connected to the microgrid with a 3-phase inverter. In **Figure 18**, 'Preated_Diesel' is set the rated power of the diesel generator. The power reference (Pref_Hydro) for the diesel generator is set to be zero. The microgrid controller can control the diesel generator's operation to be ON or OFF by the control signal from the microgrid controller (Ctrl_Diesel). The power generated

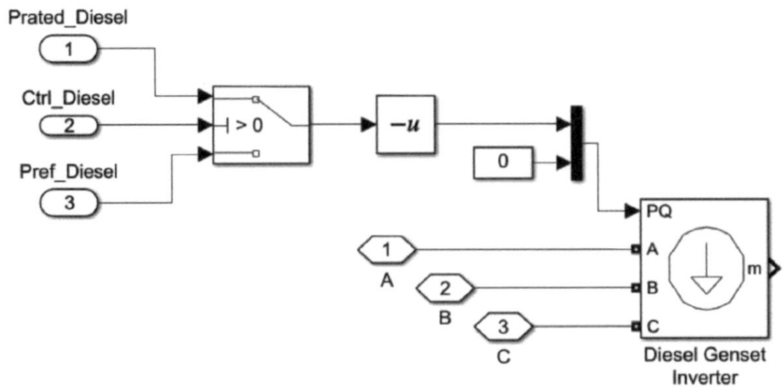

Figure 18.
Diesel generator with inverter model.

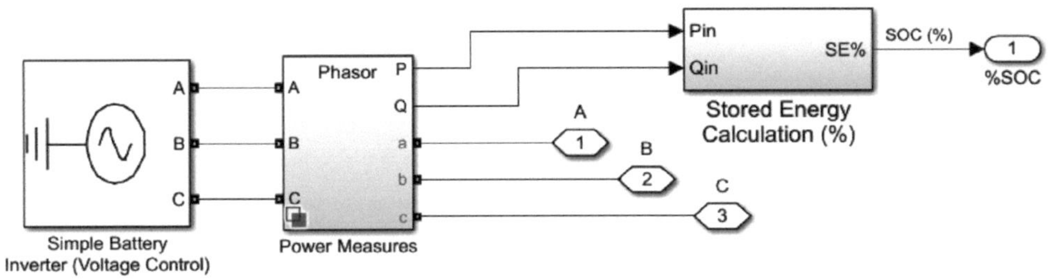

Figure 19.
Energy storage system model.

from the diesel generator can be either set to the rated power (Prated_Diesel) if Ctrl_Diesel is '1' or set to be zero (OFF) if the Ctrl_Diesel is '0'.

3.6 Energy storage system model

The energy storage system (ESS) model uses the 3-phase simple battery inverter model in MATLAB Simulink connected with the Power Measures block. Active power (P) and reactive power (Q) are measured in order to calculate the state of charge (%SOC) in the Stored Energy Calculation block. **Figure 19** shows MATLAB Simulink blocks inside the energy storage system model.

3.6.1 State of charge calculation

State of Charge (SOC) is the level of the energy storage system relative to its capacity. It can be calculated from the active power (P) and reactive power (Q), which are imported or exported from the energy storage system. The state of charge of the energy storage system can be calculated as follow:

$$SOC(t) = SOC(t-1) + \int_0^t \frac{S}{C_{ESS}} \cdot \eta \cdot dt \quad (2)$$

Where SOC(t) is the state of charge of energy storage at time t (%), SOC(t-1) is the initial battery state of charge (%), S is apparent charge/discharge power (kW), is the energy storage capacity (kWh), η is the efficiency of the energy storage system (%). t is time (h). **Figure 20** shows the MATLAB Simulink blocks for calculating the state of charge of the energy storage system.

DOI: http://dx.doi.org/10.5772/intechopen.98551

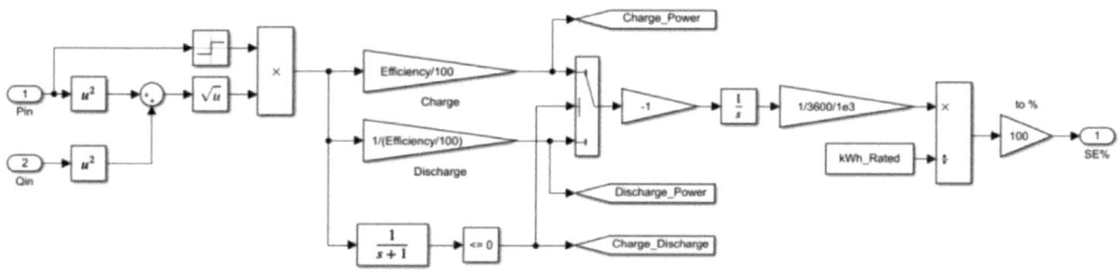

Figure 20.
State of charge of the ESS calculation model.

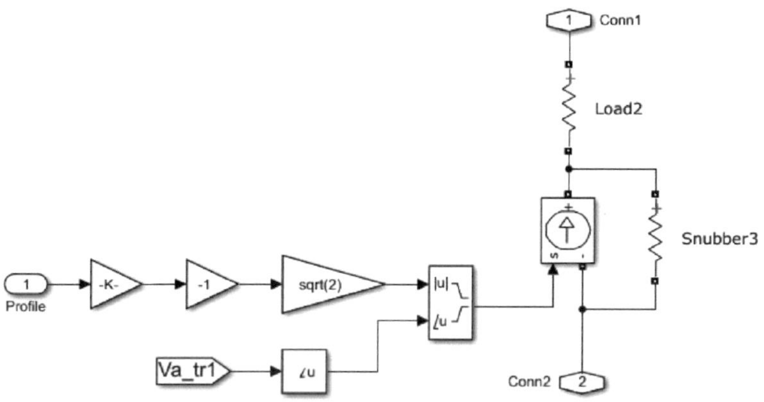

Figure 21.
Internal Simulink blocks model for each load.

3.7 Loads model

In this study, there are six loads in the off-grid microgrid. They can be categorized into two groups: (1). Non-essential loads and (2). Essential loads. Each load model consists of a current control source controlled by the current calculated from the power consumption profile imported from the MATLAB workspace. **Figure 21** shows the internal MATLAB Simulink blocks of each load.

Each load model is a single-phase load connected to the smart meter and transformer, respectively. The smart meter is a single-phase meter controlled by the microgrid controller to be connected or disconnected to the microgrid. The smart meter is also used to monitor the load's power flow by measuring the voltage and the current of the load in real-time. The inside of the smart meter model is shown in **Figure 22**. The transformer is a step-down transformer, Dyn11, which converts voltage from 22 kV on the primary side to 400 V/230 V on the secondary side, as shown in **Figure 23**.

3.7.1 Non-essential loads model

The non-essential loads are the group of residential loads. This study consists of houses1, houses2, and houses3, which are connected to the phase-a, b, and c of the secondary side of the first transformer (Tr1), respectively. Each model has its power consumption profile. The non-essential loads have lower priority than the essential loads. If load shedding is needed in the microgrid, the non-essential loads will be shed by the lowest priority from houses1, houses2, and houses3, respectively. In contrast, if load-restoration is required, the non-essential loads will be

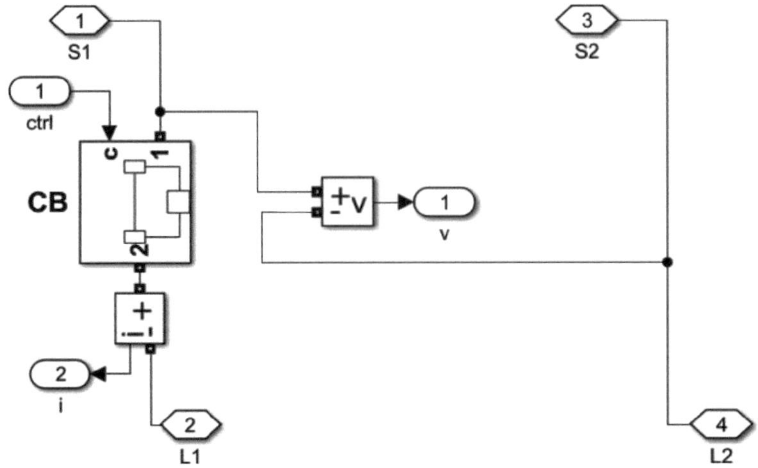

Figure 22.
Inside the smart meter model.

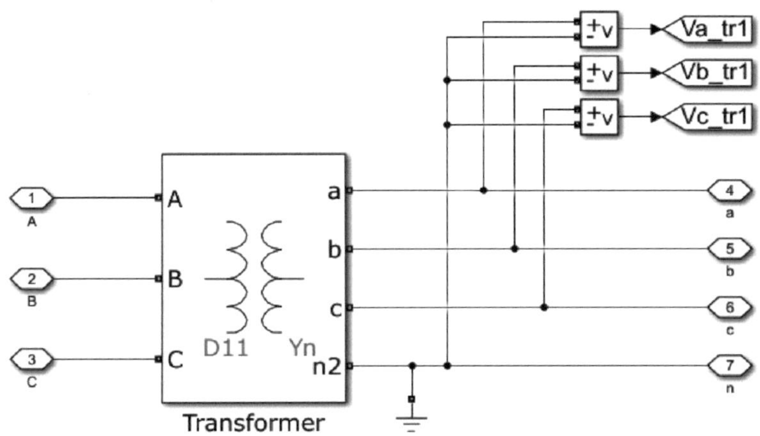

Figure 23.
Inside the load transformer model.

restored by the highest priority from houses3, houses2, and houses1, respectively. **Figure 24** shows the MATLAB Simulink model of the non-essential loads connected to the smart meter and the first transformer's secondary side.

3.7.2 Essential loads model

The essential loads have higher priority than the non-essential loads. They consist of office&school, hospital, and agriculture pumps connected to the phase-a, b, and c of the secondary side of the second transformer (Tr2), respectively. Each model has its power consumption profile. Suppose load shedding is needed in the microgrid, and all non-essential loads are already shed. In that case, the essential loads will be shed by the lowest priority from agriculture-pumps, office&school, and hospital, respectively. In contrast, if the load-restoration is required, the essential loads will be restored before the non-essential loads by the highest priority from the hospital, office&school, and agriculture pumps, respectively. **Figure 25** shows the MATLAB Simulink model of the essential loads connected to the smart meter and the second transformer's secondary side.

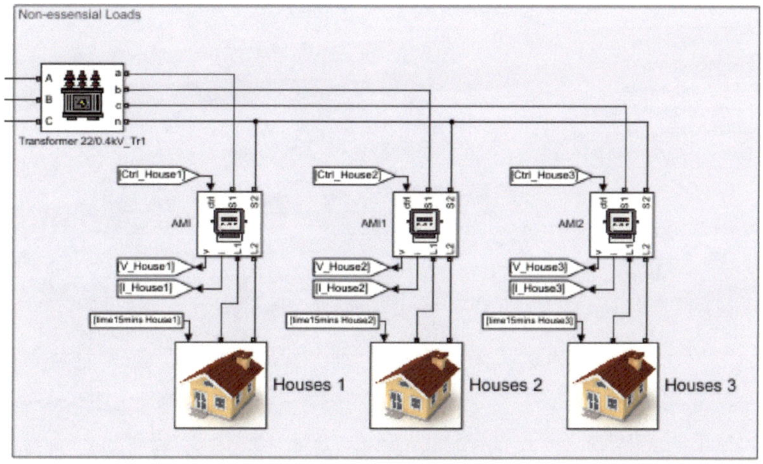

Figure 24.
Non-essential loads model.

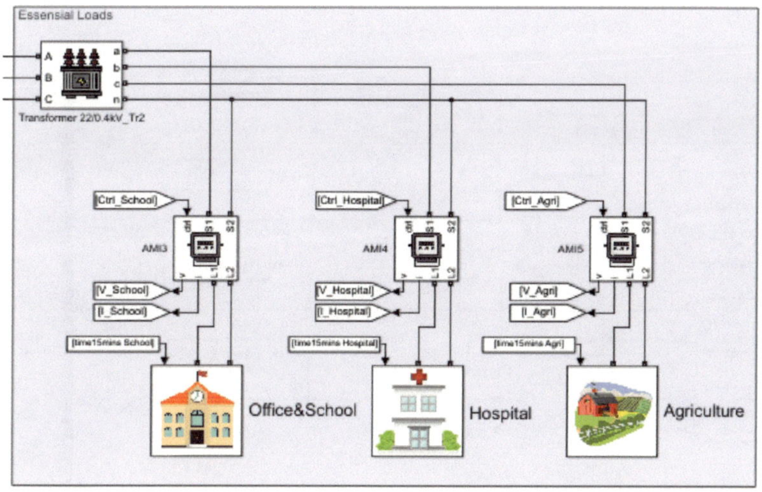

Figure 25.
Essential loads model.

3.7.3 Load profiles

Each load model is corresponding to the load consumption profile imported from the MATLAB workspace. The six load profiles are used in the simulation. All load profiles are selected from the accurate load profile from Thailand. **Figure 26** shows the daily load profile of the six loads.

3.8 Power flow measurements

The power flow among the distributed generations, energy storage system, and loads in the microgrid are calculated from the voltage and current, which are measured from each component. **Figures 27** and **28** show the power flow calculation blocks. The voltage and the current measured from the microgrid components are converted to active power (kW) to illustrate the simulation's power flow result.

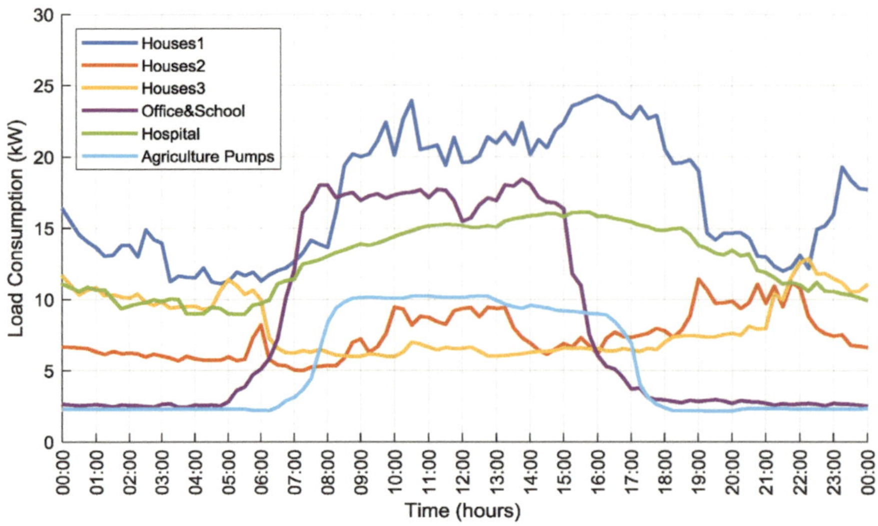

Figure 26.
Daily load profile of the 6 loads.

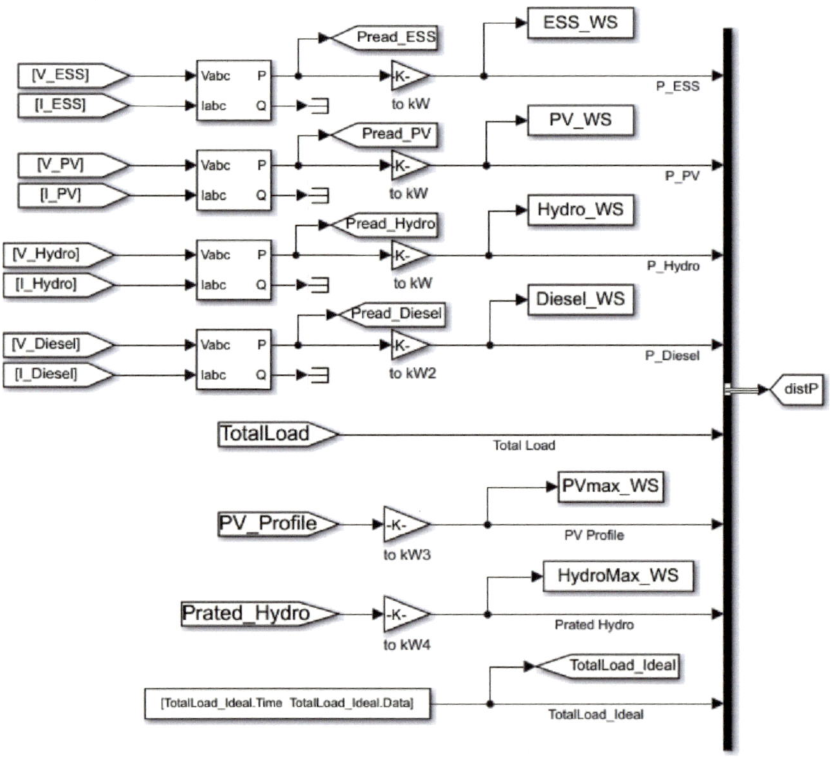

Figure 27.
Power flow calculation blocks for ESS and DGs.

3.9 Simulation result measurements

In this study, to find the optimal size of the components in the microgrid, i.e., energy storage system rated capacity, the rated power of photovoltaic,

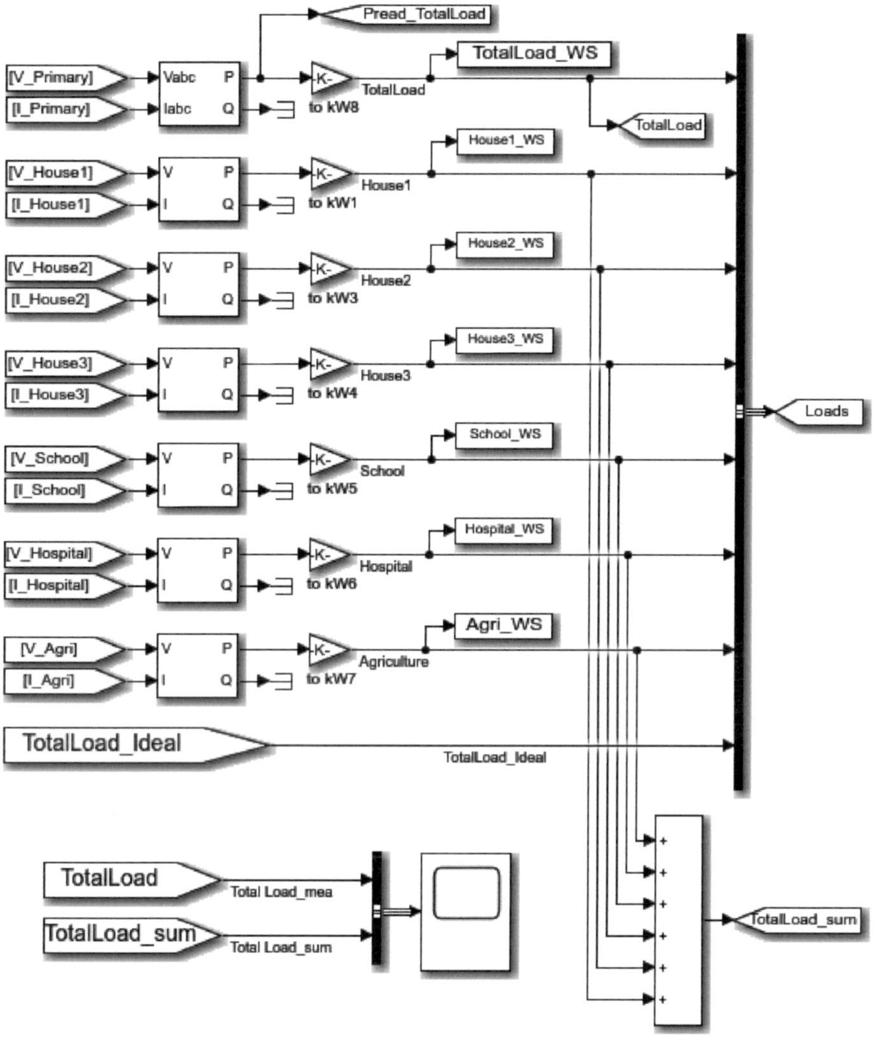

Figure 28.
Power flow calculation blocks for the loads.

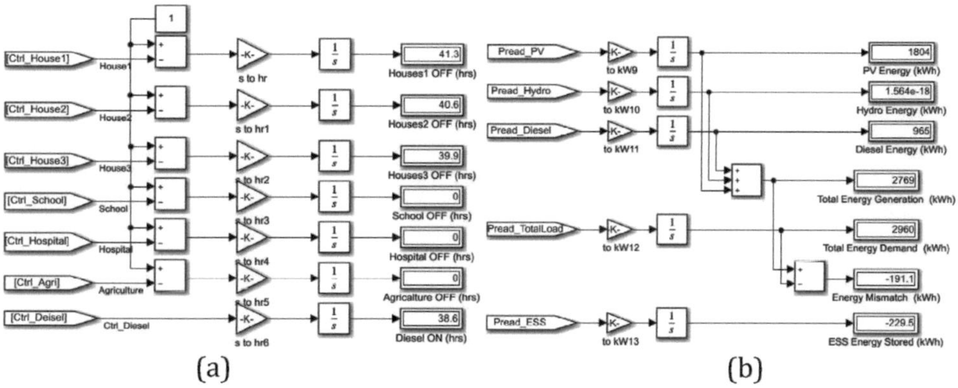

Figure 29.
(a) The blocks for calculation the duration of the power outage and ON time of the diesel generator in the microgrid simulation (b) the blocks for indication and calculation of the energy flow in the microgrid simulation.

micro-hydro, diesel generator, and the initial %SOC of the energy storage system. The power outage duration of the loads and ON time of the diesel generator are considered. The experiments for each case with different parameters will be conducted. The case in which the result has the lowest power outage and ON time of diesel generator will be considered the optimal case. MATLAB Simulink blocks for measuring and counting the power outage duration and ON time of the diesel generator have been built as shown in **Figure 29(a)**. Additionally, the energy consumed for each load and the energy supplied by each DG, the energy import/export from each simulation result's energy storage system, can be calculated and shown in the display blocks, as shown in **Figure 29(b)**.

4. Microgrid control

In this section, the concept of microgrid control is explained. The constraints of the off-grid microgrid are discussed. The structure, model, and algorithm of the Microgrid controller are shown.

4.1 Constraints

The microgrid control has the main objective to satisfy the constraints of the off-grid microgrid as follows.

Power balance constraint:

$$P_{totalload}(t) = P_{ESS}(t) + P_{PV}(t) + P_{micro-hydro}(t) + P_{diesel}(t) \tag{3}$$

Where $P_{totalload}(t)$ is the total power consumption of the loads, $P_{ESS}(t)$ is the power import/export to the microgrid from the energy storage system, $P_{PV}(t)$ is the power injecting to the microgrid from the photovoltaic system (PV), $P_{micro-hydro}(t)$ is the power injecting to the microgrid from the micro-hydro, $P_{diesel}(t)$ is the power injecting to the microgrid from the diesel generator.

Energy Storage System constraints:

ESS state of charge constraint:

$$SOC_{min} \leq SOC(t) \leq SOC_{max} \tag{4}$$

ESS power output:

$$P_{ESS}^{min} \leq P_{ESS}(t) \leq P_{ESS}^{max} \tag{5}$$

Where SOC(t) is the state of charge of the energy storage system, SOC_{min} is the lowest state of charge of the energy storage system, SOC_{max} is the maximum state of charge of the energy storage system. In this study, assuming the SOC_{max} is 90%, and the SOC_{min} is 20%.

4.2 Microgrid controller

The microgrid controller is the central controller for the off-grid microgrid. The control of microgrids is centralized. In this study, MATLAB R2020a with Simulink and Stateflow toolbox has been used to develop the microgrid controller's model and algorithm. The microgrid controller is assumed to communicate in real-time with the distributed generations' local controller, i.e., PV, micro-hydropower, diesel generator, and smart meters for the loads. **Figure 30** shows the block of the microgrid controller developed in MATLAB Simulink. The microgrid controller has five inputs and 11 outputs. The inputs are (1) state of charge of the energy storage system (SOC), real-time power supplied by (2) PV (Pread_PV), (3) micro-hydropower (Pread_Hydro), (4) diesel generator (Pread_Diesel), and (5) real-time

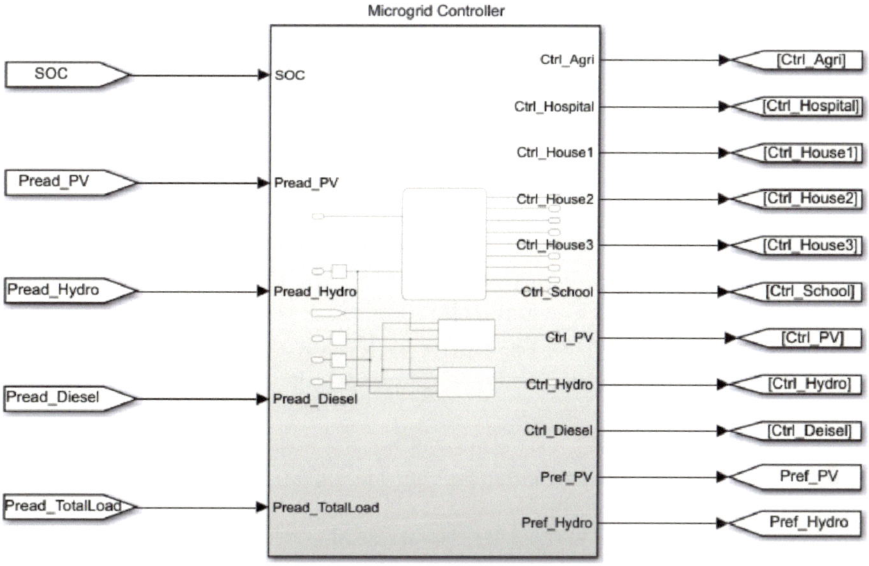

Figure 30.
Microgrid controller.

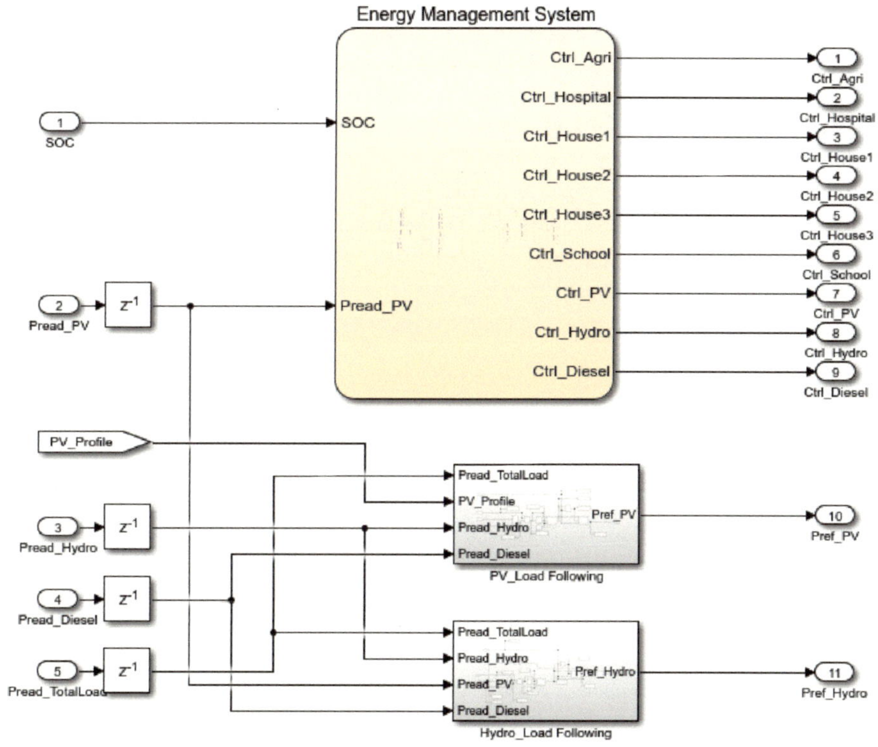

Figure 31.
Inside of the microgrid controller block.

power required by loads (Pread_TotalLoad). The output is control signals to the smart meters of the loads, i.e., (1) agriculture-pump (Ctrl_Agri), (2) hospital (Ctrl_hospital), (3) residential loads1 (Ctrl_House1), (4) residential loads2 (Ctrl_House2), (5) residential loads3 (Ctrl_House3), (6) office&school (Ctrl_School), the local controller of DGs i.e. (7) PV (Ctrl_PV), (8) micro-hydro (Ctrl_Hydro), (9) diesel generator (Ctrl_Diesel), and the power reference value for the (10) PV (Pref_PV), and (11) micro-hydro (Pref_Hydro).

4.3 Energy management system

The microgrid controller block consists of the energy management system block and blocks of load following PV and micro-hydro blocks. **Figure 31** shows the inside of the microgrid controller. The energy management system block has two inputs, i.e. SOC and Pread_PV. Also, 9 outputs for control the ON/OFF operation of the loads and DGs' smart meter. The load following block receives the measuring value of real-time power generation of PV, micro-hydro, diesel, and total real-time power required by loads to control the reference power of the local controller of PV and micro-hydro.

4.3.1 Energy management system model

The energy management system has been developed in the MATLAB Stateflow toolbox. **Figure 32** shows the energy management system function block using MATLAB Stateflow.

4.3.2 Load following models

Load following blocks are a part of the microgrid controller. The load following has the primary function to provide power reference value to the local controller, i.e., inverter of PV and the micro-hydro's inverter. The mismatch of the power demand of the loads and power generated from micro-hydro and diesel is calculated to provide a power reference value for PV inverter to inject the specific power to the microgrid. Similarly, The mismatch of the power demand of the loads and power generated from PV and diesel is calculated to provide the power reference for the

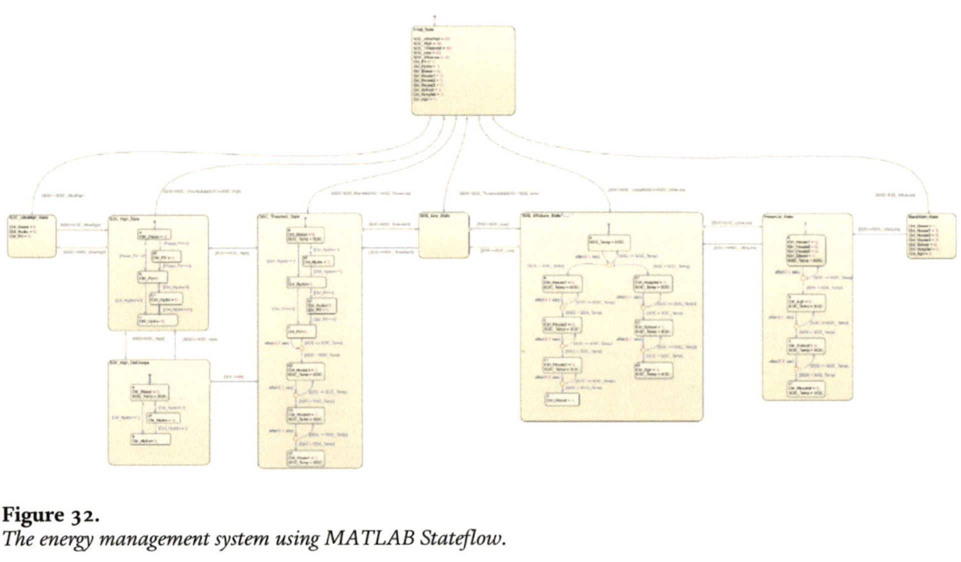

Figure 32.
The energy management system using MATLAB Stateflow.

local controller (inverter) of the micro-hydro in order to inject the specific power from the micro-hydro to the microgrid. **Figures 33** and **34** show the load following block of power reference for PV and micro-hydro, respectively.

4.3.3 Algorithm and flowchart

The algorithm for the microgrid controller in this study is the SOC-based algorithm that controls the microgrid operation related to the level of state of charge (% SOC) of the energy storage system (ESS).

- The algorithms start with initializing the state of charge levels and microgrid control signal initial parameters.
- The initializing consists of a defined SOC level, turn ON all the loads, turn OFF the diesel generator. Control the PV and micro-hydro's local controller to operate in the maximum power point mode (MPPT).
- The state of charge of the energy storage system is divided into six levels, as shown in **Table 2**.
- The microgrid controller read the data from the microgrid, i.e., total load demand, load status, power inject and import from the energy storage system, power generated from the distributed generations; PV, micro-hydro, diesel generator, voltage, and frequency of the microgrid.

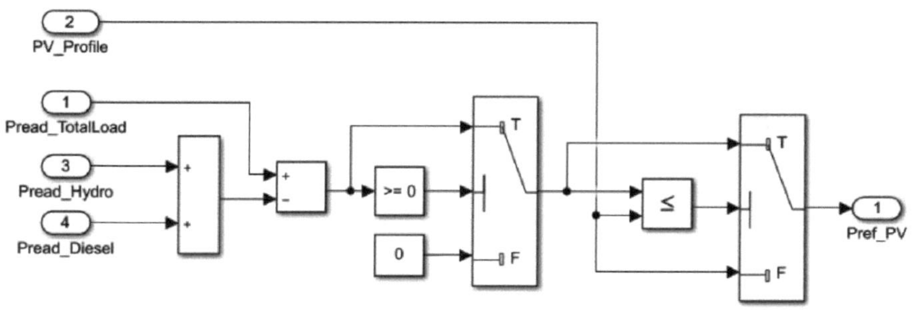

Figure 33.
Load following block of power reference for PV.

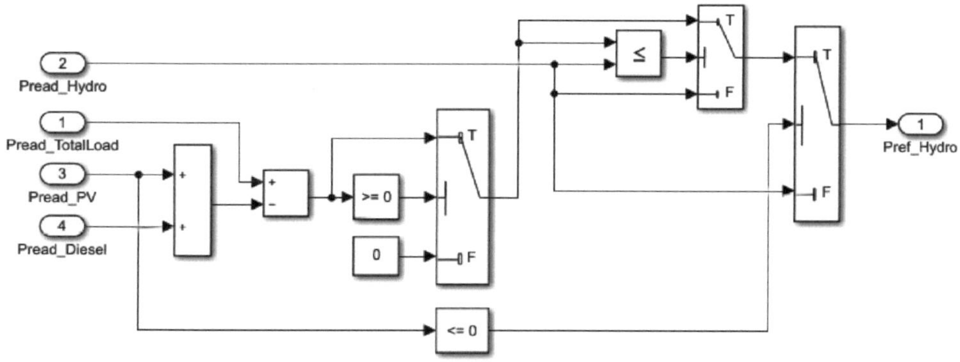

Figure 34.
Load following block of power reference for micro-hydro.

SOC levels	%SOC
SOC_UltraHigh	95
SOC_High	90
SOC_Discharge	85
SOC_Threshold	50
SOC_Low	20
SOC_UltraLow	10

Table 2.
State of charge levels of the energy storage system.

- Compare %SOC read from the ESS to the SOC level defined in the controller;
    - o If %SOC > SOC_UltraHigh ➔ Switch the control mode in the local controller of the PV and micro-hydro to be the power reference mode.
    - o If %SOC < SOC_UltraHigh and > SOC_High ➔ The controller will try maintaining the SOC level by switching the control mode of the PV to the load-following mode in which the PV is supplying the power to the microgrid as power demand from the loads. However, suppose the power generated from PV is already zero. In that case, the microgrid will switch the control mode of the micro-hydro to the load following mode instead to maintain the state of charge at this level.
    - o If %SOC < SOC_High and > SOC_Discharge ➔ The microgrid controller will switch the micro-hydro control mode back to maximum power mode.
    - o If %SOC < SOC_Discharge > SOC_Treshold ➔ The controller will turn the diesel generator OFF and switch the control mode of both micro-hydro and PV back to the maximum power mode. Then, if the SOC is increasing, all the loads will be gradually restored.
    - o If %SOC < SOC_Threshold and > SOC_Low ➔ Nothing change in the signal from the microgrid central controller to the local controller.
    - o If %SOC < SOC_Low and > SOC_UltraLow ➔ The controller will check whether the SOC is increasing or decreasing. If the SOC is increasing, the essential loads will be restored. If the SOC is decreasing, the Non-essential loads will be shed. Finally, if all non-essential loads are shed but the SOC is still decreasing, the diesel generator will be turned ON.
    - o If %SOC < UltraLow ➔ The microgrid controller will keep the diesel generator inject the power to the microgrid. If the SOC is still decreasing, the essential loads will be shed until the is no load in the system. The diesel generator will charge the ESS until the %SOC is changed to another level.
- After finished updating the outputs, the program will continuously return to read the inputs and compare the %SOC to the SOC level again.
- All the processes will continue repeat all the time to operate the microgrid.

DOI: http://dx.doi.org/10.5772/intechopen.98551

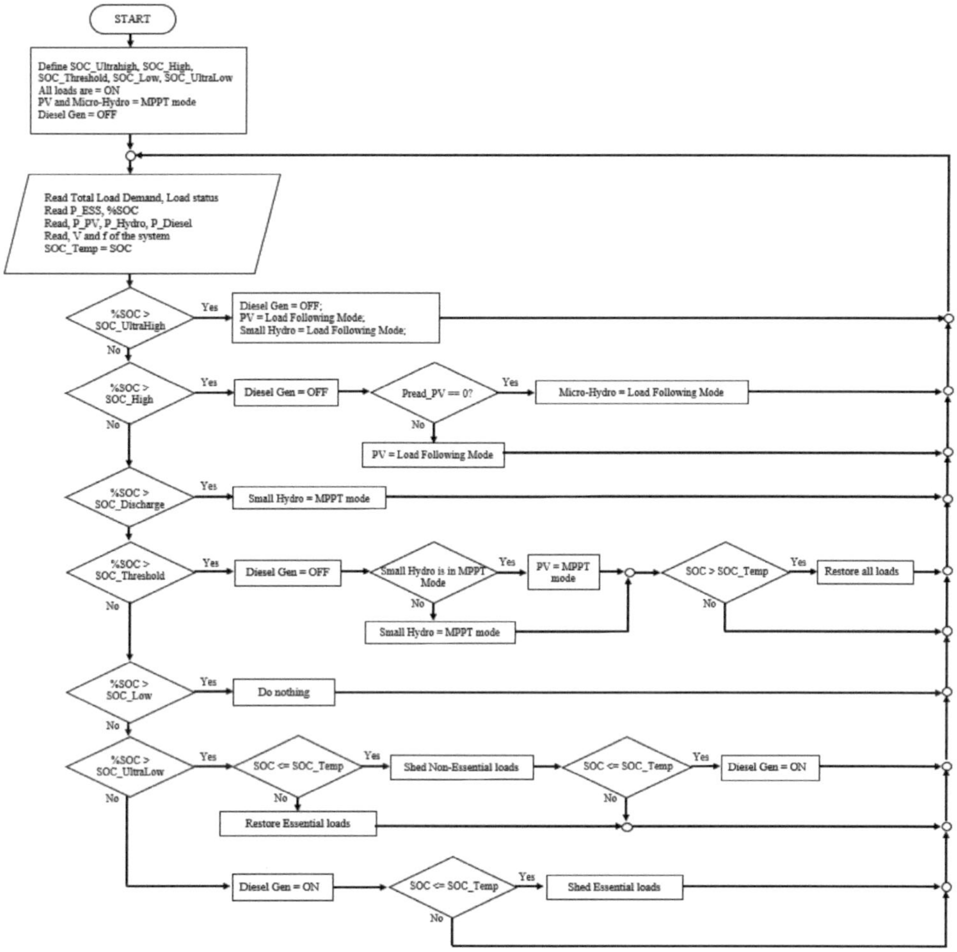

Figure 35.
Flowchart of the microgrid control algorithm.

The flowchart of the energy management system in the microgrid controller can be shown in **Figure 35**.

5. Results and discussions

5.1 Optimal sizing for the microgrid based on the proposed controller

To find the optimal size for the microgrid distributed generations and Energy storage capacity. The microgrid is simulated in various sizes of each element. The microgrid's optimal size based on the proposed control algorithm is evaluated by the duration of the power outage and diesel operation (ON) in hours, as shown in **Figure 36**. The simulation is conducted for 72 hours (3 days), and the initial SOC for the first day is fixed at 55%. The simulation can be separated into 24 cases for each test. To find the suitable power rating of the diesel generator, whether 25 kW or 50 kW and the low-level of state of charge (SOC_Low) of the energy storage system, the test can be separated into four tests:

1. The diesel generator is 25 kW with SOC_Low = 30%.
2. The diesel generator is 50 kW with SOC_Low = 30%.
3. The diesel generator is 25 kW with SOC_Low = 20%.
4. The diesel generator is 50 kW with SOC_Low = 20%.

The simulation results will be illustrated in the four graphs, as shown in **Figure 36**.

a. Power flow.

The power flow graph shows the power flow in kilowatt (kW) of the energy storage system (ESS), all of the distributed generation (DGs) in the microgrid, i.e., photovoltaic (PV), micro-hydro plant, diesel generator, and loads. The positive results mean the power is injected from the ESS, DGs to the microgrid. In contrast, the negative results mean that the power is imported to the ESS.

b. Energy storage system state of charge.

Shows the state of charge (SOC) in percentage (%) of the energy storage system in the microgrid.

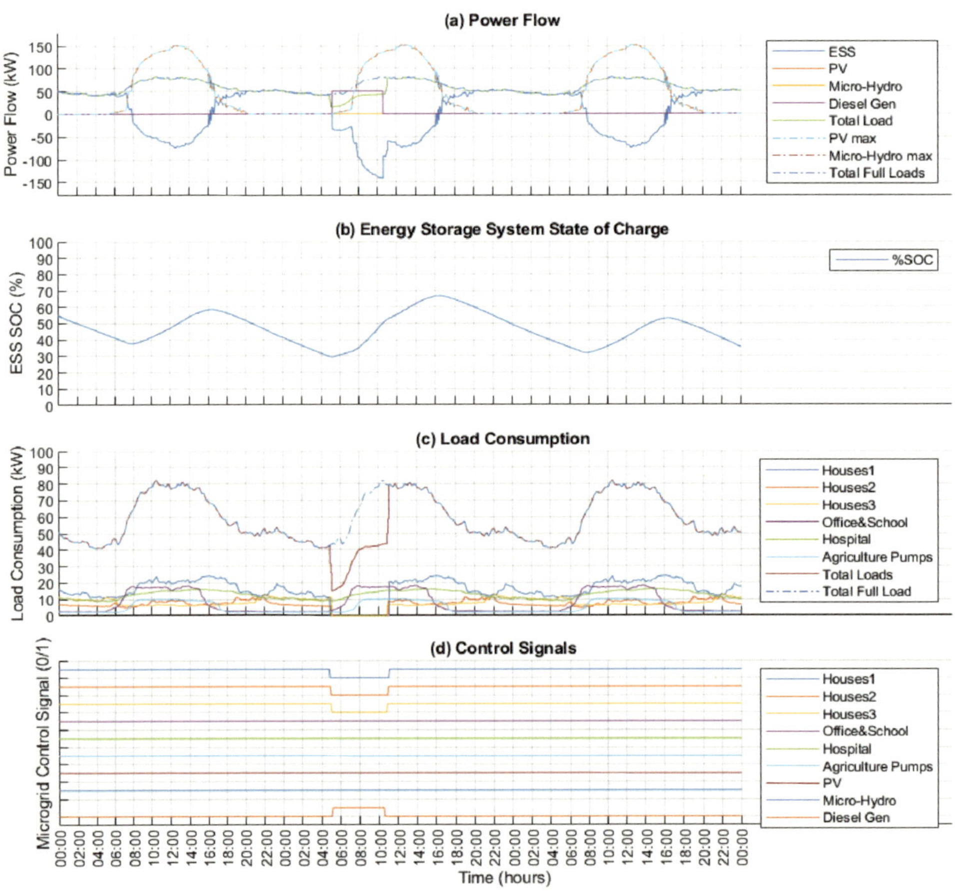

Figure 36.
The duration of power outage and diesel generator in turned on are counted in each simulation case.

c. Load Consumption.

Shows the load profile of the total load and also for each load i.e. Houses1, Houses2, Houses3, Offices&School, Hospital, Agriculture Pump.

d. Control signal.

e. Shows each control signal of the microgrid controller which send to the local controller e.g. PV inverter, smart meters etc. The signal can be 1 (ON) or 0 (OFF).

For each test, the power rating of micro-hydro is varied from 0, 25, 50 kW. The maximum power rating of photovoltaic (PV) is varied from 100 and 200 kWp. The daily PV power generation profile is varied from Maximum sunlight profile (Max) and Minimum sunlight profile (Min). The energy storage system rated capacity is varied from 1000 to 2000 kWh. **Table 3** shows the duration of the power outage and diesel generator ON in hours simulated in 72 hours in the MATLAB Simulink off-grid microgrid model using the proposed control algorithm in the microgrid controller. It can be considered that there is no power outage, and the diesel generator is turned OFF all the time when micro-hydro generates the maximum power at 50 kW. For the worst-case scenario, there are 8 cases, i.e., case 1, 2, 7, 8, 13, 14, 19, 20 which the micro-hydro is unavailable (micro-hydro is zero). It can be seen that the shortest duration of the power outage and the diesel generator is ON is occur in case 19 with test 2 as shown in the red rectangle in **Figure 37(a)** and **(b)**.

Therefore, it can be summarized that the optimal parameter for the off-grid microgrid using the proposed algorithm to have the shortest duration of power outage and duration of the diesel generator is turned ON is as follows:

- nergy Storage System rated capacity: 2000 kWh.
- Photovoltaic (PV) power rating: 200 kWp.
- Micro-hydro rated power: 50 kW.
- Diesel Generator rated power: 50 kW.
- The SOC_Low is set at 30%.

5.2 Response of the microgrid controller

In this section, the microgrid model is simulated in 24 hours using the proposed microgrid controller algorithm. In order to see the microgrid controller's response for each situation, the eight scenarios can be shown in **Table 4**. According to the results from 5.1, the optimal parameter of the microgrid has been investigated. The energy storage system rated capacity is fixed to 2000 kWh, 100 kW. The photovoltaic (PV) rated power is fixed to 200 kWp. The Diesel generator-rated power is fixed to 50 kW. In order to see the response of the proposed microgrid controller, the initial SOC will be varied from 80 to 25%, the PV generation profile will be varied from Max and Min, the micro-hydro will be varied from 50 kW (max) and 0 kW (min).

The details of the simulation for each scenario can be explained as follows:

Case	Variables				Outage (hours)				Diesel ON (hours)			
	Hydro (kW)	PV (kW)	PV (Case)	ESS (kWh)	Test1	Test2	Test3	Test4	Test1	Test2	Test3	Test4
1	0	100	Max	1000	43.5	44.4	40.6	33.7	41.5	22.2	38.6	27.1
2	0	100	Min	1000	66.4	66.4	64.4	64.4	66.2	35.2	64.2	36.9
3	25	100	Max	1000	18.6	18.6	14.4	13.2	0	0	5	3.8
4	25	100	Min	1000	48.2	49.6	46.9	44.5	19.8	9.5	20.4	13
5	50	100	Max	1000	0	0	0	0	0	0	0	0
6	50	100	Min	1000	7.3	7.3	8.6	8.6	0	0	0	0
7	0	200	Max	1000	18.1	15	14.2	7.6	16.3	13.2	13	7
8	0	200	Min	1000	66.4	66.4	64.4	64.4	54.4	27.4	49.3	30.7
9	25	200	Max	1000	0	0	0	0	0	0	0	0
10	25	200	Min	1000	51.7	51.7	42.3	40.9	0	0	10.3	5.9
11	50	200	Max	1000	0	0	0	0	0	0	0	0
12	50	200	Min	1000	0	0	0	0	0	0	0	0
13	0	100	Max	2000	38.0	32.2	38.5	27.6	36.5	23.8	33.6	26.4
14	0	100	Min	2000	62	62	59.2	59.2	61.8	32.2	59	33.7
15	25	100	Max	2000	14.3	14.3	0	0	0	0	0	0
16	25	100	Min	2000	42.4	40.5	50.6	50.6	13.2	7.7	0	0
17	50	100	Max	2000	0	0	0	0	0	0	0	0
18	50	100	Min	2000	0	2.4	0	0	0	0	0	0
19	0	200	Max	2000	12.1	6	10.3	8.7	10.9	5.4	9.7	8.1
20	0	200	Min	2000	61.5	61.5	58.1	58.1	61.3	26.8	57.9	27.2
21	25	200	Max	2000	0	0	0	0	0	0	0	0
22	25	200	Min	2000	53.7	53.7	46.6	46.6	0	0	0	0
23	50	200	Max	2000	0	0	0	0	0	0	0	0
24	50	200	Min	2000	0	0	0	0	0	0	0	0

Table 3.
The number of hours of power outage and diesel generator ON in 72 hours for 4 tests in 24 cases.

5.2.1 Scenario 1

Scenario 1 illustrates the day's microgrid operation with high initial SOC and maximum power generated from PV. At the beginning of the day (00:00), the energy storage system has 80% of SOC, and the PV generates power using maximum sunlight profile. The micro-hydro is fully generating power at 50 kW constantly throughout the day. All the essential and non-essential loads are ON. PV and micro-hydro are operating in maximum power mode (MPPT). The diesel generator is OFF, as shown in **Figure 38(d)**.

In **Figure 38**, the simulation result shows that the total demand from the loads is a little bit less than the total power generation, which mainly from the micro-hydro. ESS is charging slowly from 00:00 until the sunlight comes at around 06:00. The excess energy generated from PV and micro-hydro supplied to the loads is charged to ESS.

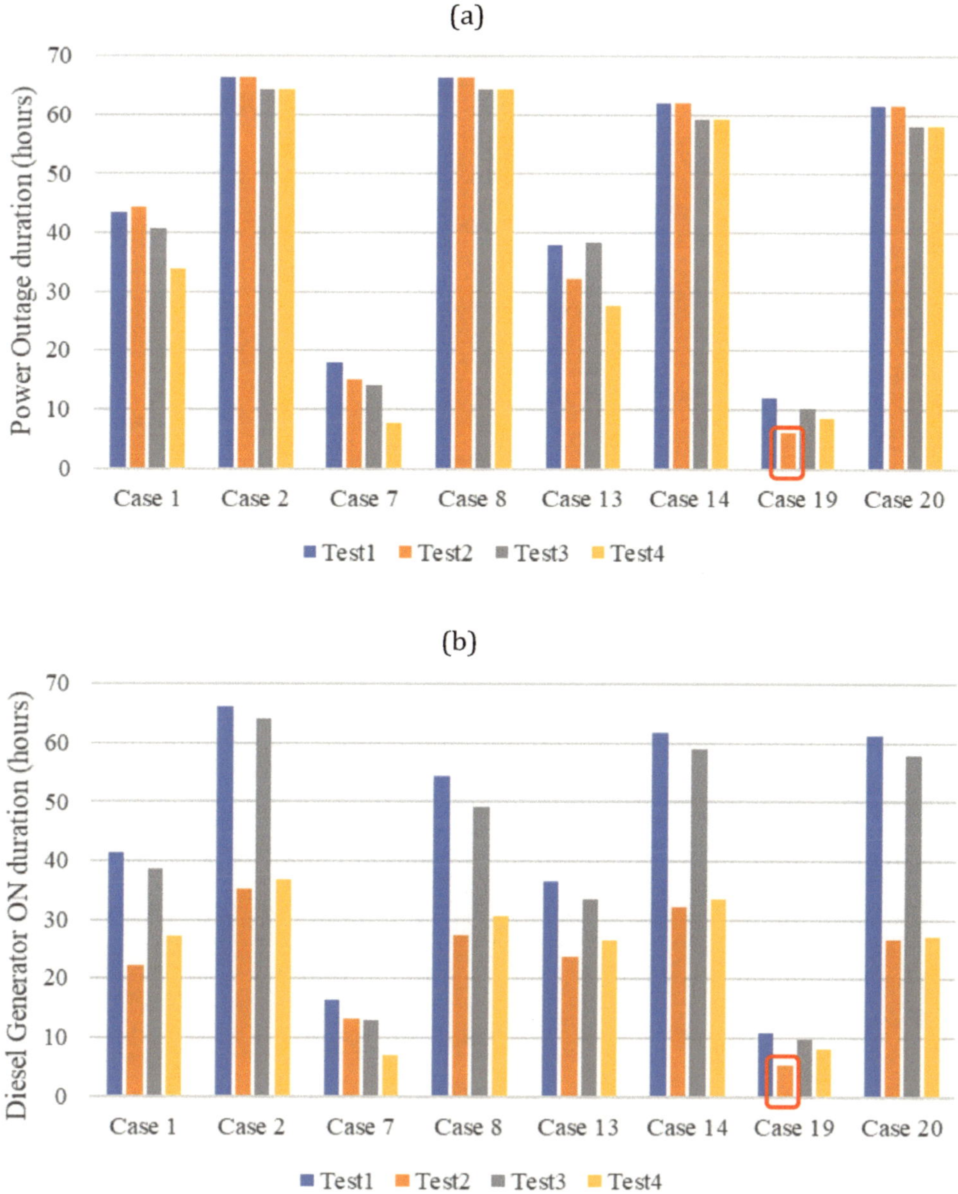

Figure 37.
The result comparison only the case when micro-hydro is unavailable (a) power outage duration (hours), (b) diesel generator ON (hours).

ESS is charged until the %SOC is reached SOC_high, which is 90%. The microgrid controller changes the logic signal from '1' to '0' to tell the local controller of the PV to change the operation mode from 'MPPT mode' to 'power reference mode' at around 09:30, as shown in **Figure 38(a)**.

In the power reference mode, to maintain the %SOC of ESS, the microgrid controller calculates the power reference value to the PV to meet the loads' exceeding demand. The PV injects the power according to the power reference sending from the microgrid controller to the microgrid to supply the loads demand. The % SOC of the energy storage system is maintained at 90% (SOC_High) throughout the day, as shown in **Figure 38(b)**.

Microgrid Parameters	Scenarios							
	1	2	3	4	5	6	7	8
ESS Capacity (kWh)	2000	2000	2000	2000	2000	2000	2000	2000
Initial SOC (%)	80	80	35	35	80	80	35	35
PV power (kWp)	200	200	200	200	200	200	200	200
Sunlight Profile	Max	Min	Max	Min	Max	Min	Max	Min
Micro-Hydro power (kW)	50	50	50	50	0	0	0	0
Diesel Gen power (kW)	50	50	50	50	50	50	50	50

Table 4.
The parameters for the microgrid simulation for each scenario.

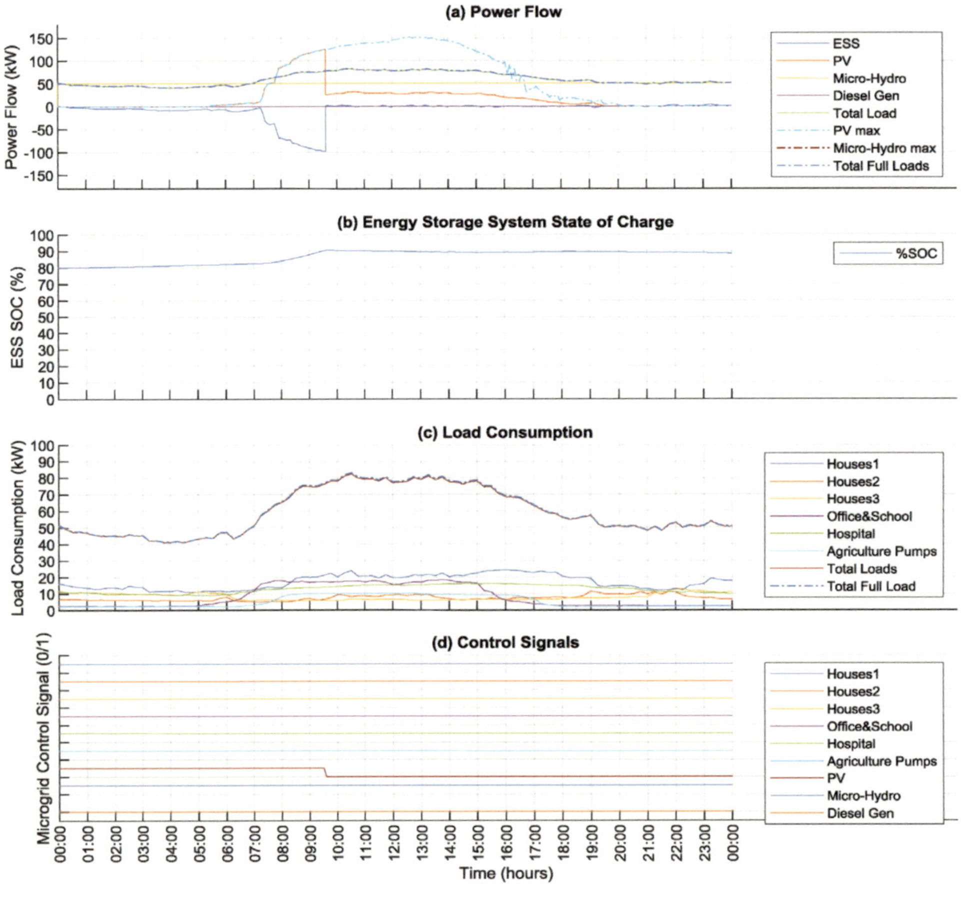

Figure 38.
The simulation result for scenario 1 (a) power flow, (b) ESS SOC, (c) load consumption, (d) control signal.

5.2.2 Scenario 2

Scenario 2 illustrates the microgrid operation in the day with high initial SOC and minimum sunlight. At the beginning of the day (00:00), the energy storage system has 80% of SOC. The PV generates power using a minimum sunlight profile. The micro-hydro is fully generating power at 50 kW constantly throughout the day. All the essential and non-essential loads are ON. PV and micro-hydro are operating in maximum power mode (MPPT). The diesel generator is OFF, as shown in **Figure 39(d)**.

In **Figure 39**, the simulation result shows that the power generation from the PV and micro-hydro is nearly met the total demand throughout the day. Thus, the % SOC of the energy storage system is constantly maintained at around 80% throughout the day. There is no changing signal from the microgrid controller, as shown in **Figure 39(b)** and **(d)**, respectively. However, The SOC of ESS has not reached the maximum level as compared to Scenario 1.

5.2.3 Scenario 3

Scenario 3 illustrates the microgrid operation in the day with low initial SOC and maximum sunlight. At the beginning of the day (00:00), the energy storage system has 35% SOC. The PV generates power using a maximum sunlight profile. The micro-hydro is fully generating power at 50 kW constantly throughout the day. All the essential and non-essential loads are ON. PV and micro-hydro are operating in maximum power mode (MPPT). The diesel generator is OFF, as shown in **Figure 40(d)**.

In **Figure 40**, the simulation result shows that the total demand from the loads is a little bit less than the total power generation, mainly from the micro-hydro. ESS is charging slowly from the beginning of the day. The PV is generating the power to

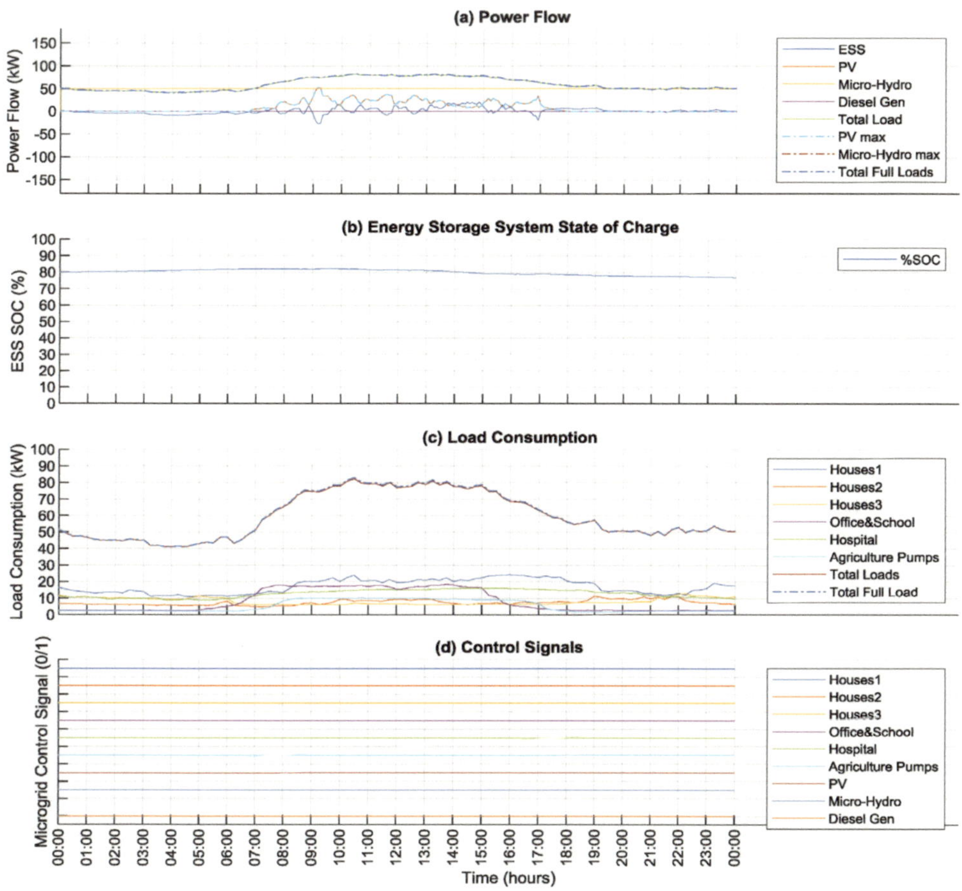

Figure 39.
The simulation result for scenario 2 (a) power flow, (b) ESS SOC, (c) load consumption, (d) control signal.

the microgrid, and ESS is also charging from 06:00 to 20:00. The state of charge (%SOC) of ESS is rising from 35–70% until the end of the day, as shown in **Figure 40(a)** and **(b)**, respectively. However, The SOC of ESS has not reached the maximum level as compared to Scenario 1.

5.2.4 Scenario 4

Scenario 4 illustrates the microgrid operation in the day with low initial SOC and minimum sunlight. At the beginning of the day (00:00), the energy storage system has 35% SOC. The PV generates power using a minimum sunlight profile. The micro-hydro is fully generating power at 50 kW constantly throughout the day. All the essential and non-essential loads are ON. PV and micro-hydro are operating in maximum power mode (MPPT). The diesel generator is OFF, as shown in **Figure 41(d)**.

In **Figure 41**, the simulation result shows that the power generation from the PV and micro-hydro is nearly met the total loads demand throughout the day. Thus, the %SOC of the energy storage system is constantly maintained at around 32–35% throughout the day. There is no changing signal from the microgrid controller, as shown in **Figure 41(b)** and **(d)**, respectively. This scenario has a similar operation result with Scenario 2, both Scenario 2 and 4 also maintain the SOC level throughout

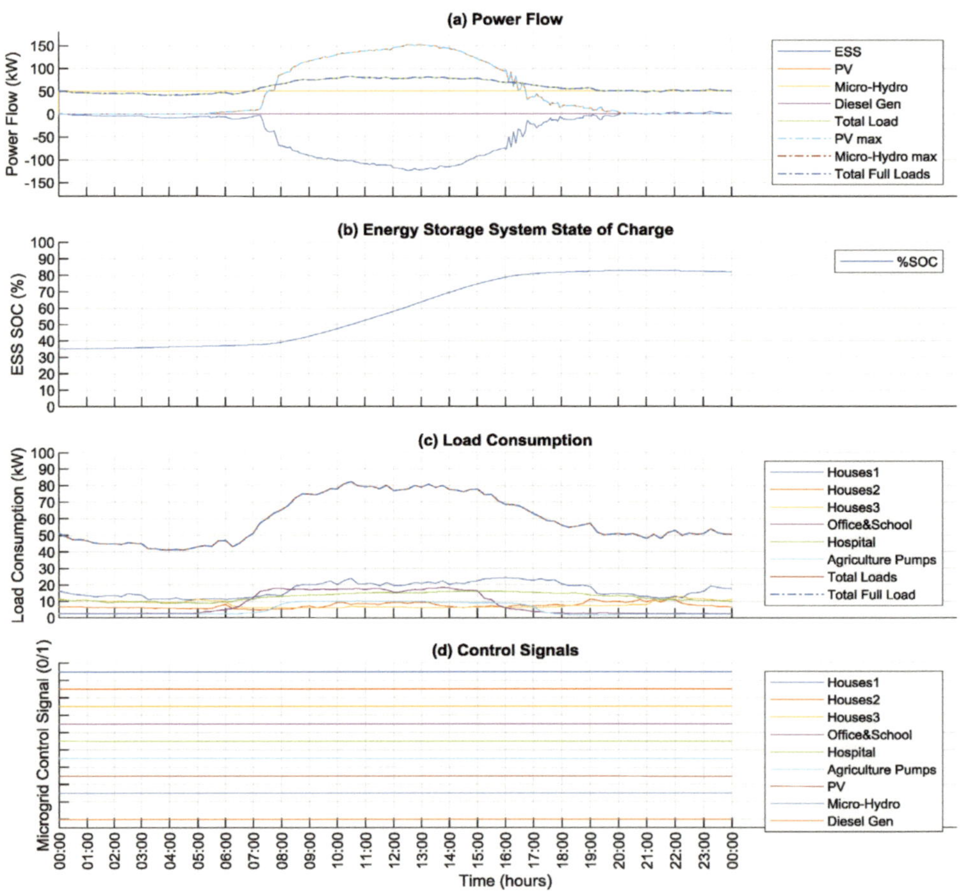

Figure 40.
The simulation result for scenario 3 (a) power flow, (b) ESS SOC, (c) load consumption, (d) control signal.

the day, but the SOC level is different. In scenario 2, SOC is maintained at a high level (80%), but in scenario 4, SOC is maintained at a low level (35%).

5.2.5 Scenario 5

Scenario 5 illustrates the microgrid operation in the day with high initial SOC and maximum sunlight, but no power is generated from the micro-hydro. At the beginning of the day (00:00), the energy storage system has 80% of SOC. The PV generates power using a maximum sunlight profile, assuming no power is generated from the micro-hydro due to unavailable season. All the essential and non-essential loads are ON. PV is operating in maximum power mode (MPPT). The diesel generator is OFF, as shown in **Figure 42(d)**.

In **Figure 42**, the simulation result shows that the energy storage is discharged to supplied energy to the microgrid from 00:00 to around 06:00 because there is no power either from PV or micro-hydro. From around 06:00 to 16:30, when sunlight comes, ESS is charged because total power generation is more than total power demand. The ESS is discharged from 16:30 until the end of the day because the power generation is less than the power demand, as shown in **Figure 42(a)** and **(b)**.

There is no change in the control signal from the microgrid controller. The microgrid is very dependent on energy storage for efficient operation but does not require a diesel generator, as shown in **Figure 42(d)**. Additionally, the SOC of ESS is maintained between 60–80% throughout the day.

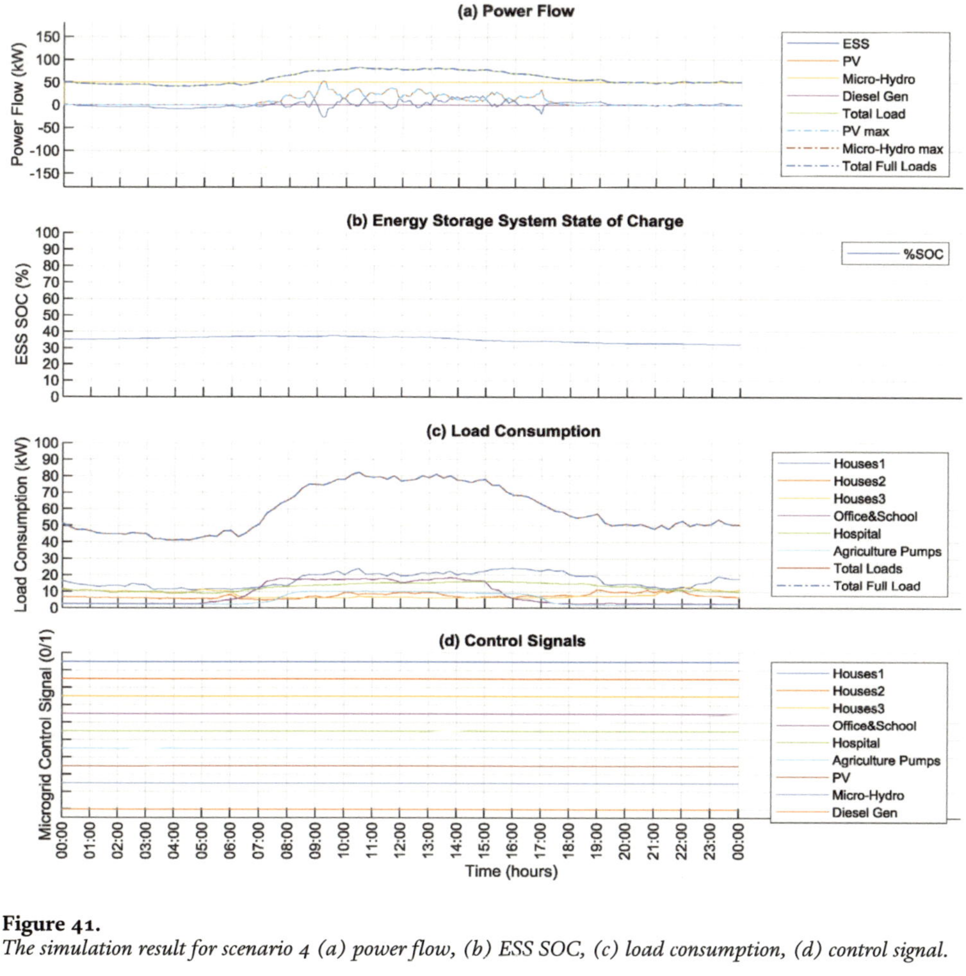

Figure 41.
The simulation result for scenario 4 (a) power flow, (b) ESS SOC, (c) load consumption, (d) control signal.

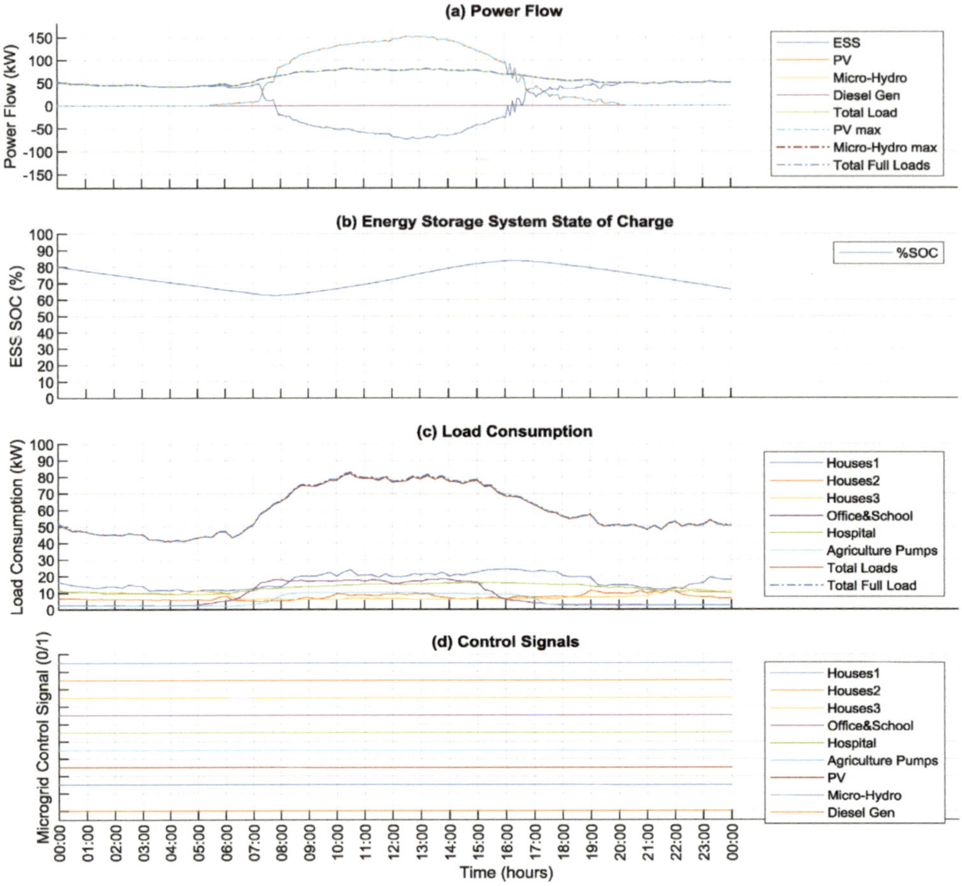

Figure 42.
The simulation result for scenario 5 (a) power flow, (b) ESS SOC, (c) load consumption, (d) control signal.

5.2.6 Scenario 6

Scenario 6 illustrates the microgrid operation in the day with high initial SOC and minimum sunlight, but there is no power generated from the micro-hydro. At the beginning of the day (00:00), the energy storage system has 80% of SOC. The PV generates power using a minimum sunlight profile, assuming no power is generated from the micro-hydro due to unavailable season. All the essential and non-essential loads are ON. PV is operating in maximum power mode (MPPT). The diesel generator is OFF, as shown in **Figure 43(d)**.

In **Figure 43**, the simulation result shows that when there is no power from the micro-hydro and the power from PV is not enough to meet the energy demand, the ESS is discharged from the beginning of the day to supply the energy microgrid.

The ESS is continuously discharged from the beginning of the day (00:00) until the state of charge of the energy storage system is reaching the SOC_min (30%) at around 19:00, the microgrid controller gradually sheds the non-essential loads from the microgrid to reduce the total energy demand from the loads as shown in **Figure 43**.

However, after turned off all non-essential loads already, if the %SOC is still decreasing, the microgrid controller sends a signal to turn on the diesel generator to supply the power to the essential loads, as shown in **Figure 43** at 19:00.

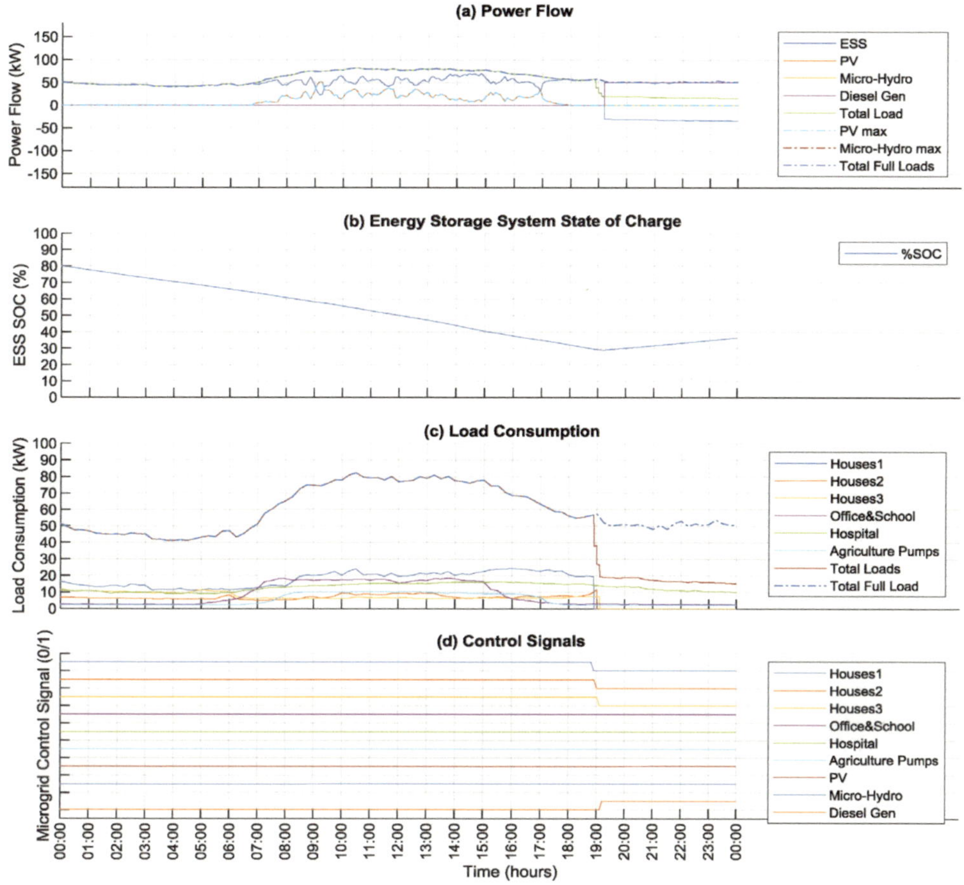

Figure 43.
The simulation result for scenario 6 (a) power flow, (b) ESS SOC, (c) load consumption, (d) control signal.

From 19:00 to the end of the day, the diesel generator runs to generate the maximum power to supply the microgrid's essential loads. The exceed energy is charging the energy storage system.

5.2.7 Scenario 7

Scenario 7 illustrates the microgrid operation in the day with low initial SOC and maximum sunlight, but no power is generated from the micro-hydro. At the beginning of the day (00:00), the energy storage system has 35% SOC. The PV generates power using a maximum sunlight profile, assuming no power is generated from the micro-hydro due to unavailable season. All the essential and non-essential loads are ON. PV is operating in maximum power mode (MPPT). The diesel generator is OFF, as shown in **Figure 44(d)**.

In **Figure 44**, the simulation result shows that the energy storage system is discharged to supply the energy to the microgrid. There is no energy generated from PV and micro-hydro at the beginning of the day (00:00). The ESS is discharged until the %SOC is below the SOC_Low (30%) at 02:20, the microgrid controller gradually sheds all the non-essential loads. Finally, the diesel generator is turned on to supply the power to the essential loads in the microgrid, as shown in **Figure 44**.

At 02:30, the ESS is charging by the diesel generator's excess energy until %SOC reaches the SOC_Threshold level (50%) at 10:00. The microgrid controller turns off the diesel generator, and then %SOC continues to increase. The microgrid controller restores the non-essential loads to the microgrid. PV starts to generate the power to the microgrid when the sunlight comes at 06:00. The energy generated from PV is higher than the energy demand from the loads. Thus, the ESS is continuing charging.

At around 16:30, the power generated from PV is decreased. The power demand from the loads is higher than total power generation in the microgrid. Thus, the ESS is discharged until the end of the day, as shown in **Figure 44(a)** and **(b)**.

5.2.8 Scenario 8

Scenario 8 illustrates the microgrid operation in the day with low initial SOC and minimum sunlight, and also, there is no power generated from the micro-hydro. At the beginning of the day (00:00), the energy storage system has 35% SOC. The PV generates power using a minimum sunlight profile, assuming no power is generated from the micro-hydro due to unavailable season. All the essential and non-essential loads are ON. PV is operating in maximum power mode (MPPT). The diesel generator is OFF, as shown in **Figure 45(d)**.

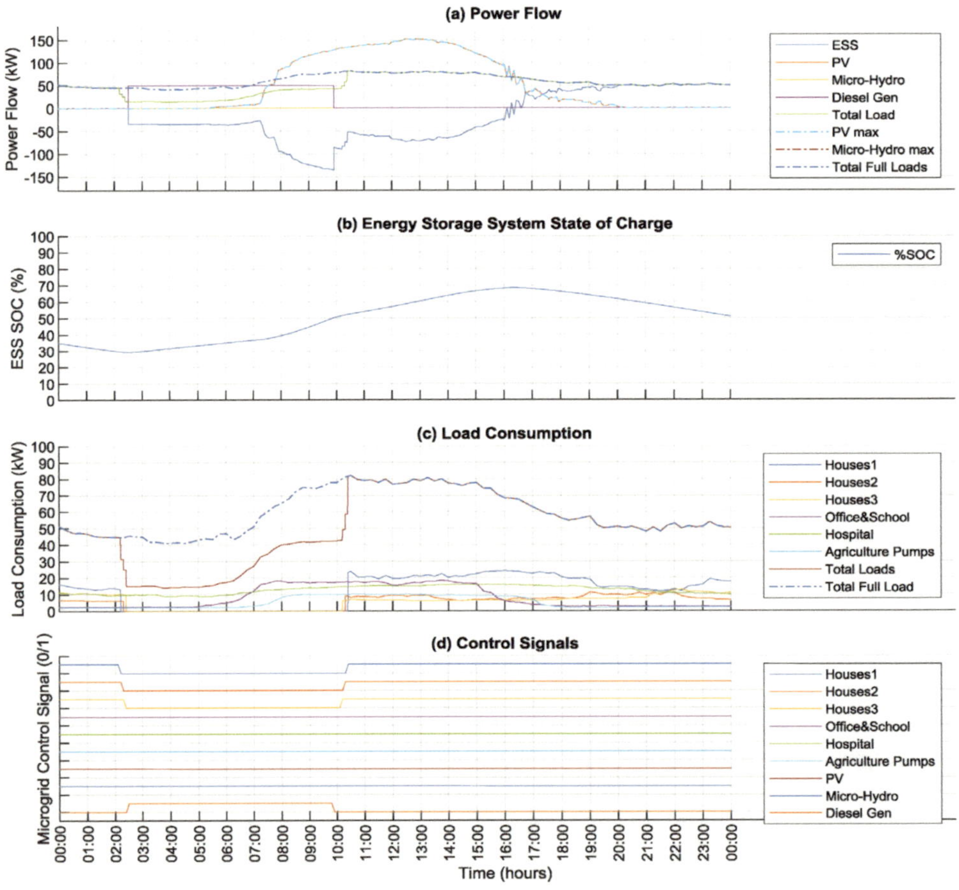

Figure 44.
The simulation result for scenario 7 (a) power flow, (b) ESS SOC, (c) load consumption, (d) control signal.

In **Figure 45**, the simulation result shows that the energy storage system is discharged to supply the energy to the microgrid due to there is no energy generated from PV and micro-hydro at the beginning of the day (00:00). The ESS is discharging until the %SOC is below the SOC_Low (30%), the microgrid controller sheds the non-essential loads. Finally, the diesel generator is turned on to supply the power to the essential loads in the microgrid at 02:20, as shown in **Figure 45**.

The ESS is charged by the diesel generator's energy and the PV until %SOC reaches SOC_Threshold level (50%). The microgrid controller turns off the diesel generator, but after the diesel generator is turned off, the %SOC is not increasing. Thus, the non-essential loads are not restored. After that, the ESS is discharging to supply the power to the essential loads from 16:30 until the end of the day, as shown in **Figure 45(b)**.

It can be considered that this scenario is the weakest mode of operation for the microgrid as both load shedding and diesel operation are required. The non-essential loads are disconnected from the microgrid from 02:20 until the end of the day. It is almost the day that the non-essential load has the blackout. Furthermore, the diesel generator must be operated for up to 14 hours to maintain the microgrid and supply power to the essential loads.

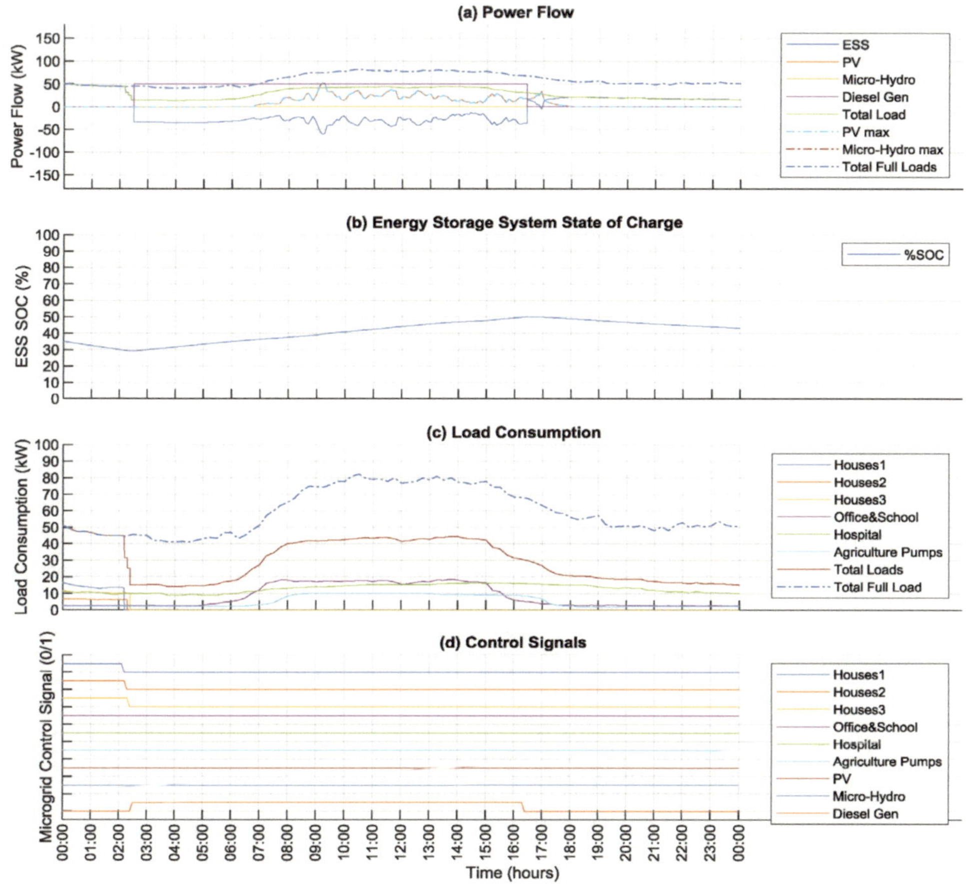

Figure 45.
The simulation result for scenario 8 (a) power flow, (b) ESS SOC, (c) load consumption, (d) control signal.

5.2.9 Summary

From the results of the simulation in scenario 1–8, it can be summarized as follows:

- The microgrid will have a normal operation. All loads are accessed to the power without any outage. The diesel generator is not required to operate if the micro-hydro is available and generates power to the microgrid at a maximum power rating of 50 kW.

- The non-essential loads will face some power outage. The diesel generator must run to maintain the microgrid and supply the essential loads when the micro-hydro is unavailable. Additionally, on the day that PV also generates power with minimum sunlight profile, the power outage duration and diesel operation time will be longer.

6. Conclusions

6.1 Main outcomes

This study has attempted to investigate the operation of the microgrid controller and the responses in various scenarios of the off-grid microgrid components, i.e., the state of charge of the energy storage system, the power generation from the distributed generations, and the essential and non-essential loads. From the results of the study, it can be concluded as follows:

- The size of the components in the off-grid microgrid is essential. The size of the distributed generators must be considered to meet the demand of the energy from the loads.

- The renewable energy resources (RES) in the microgrid, i.e., photovoltaic (PV) and micro-hydro depend on the weather condition. Solar PV depends on the sunlight of the day, micro-hydro power depends on the water level in the weir and the season. Thus, the power from these RES is intermittent. Energy storage plays a crucial role in absorbing energy when the generation is higher than demand and supplies the power when the generation is less than demand.

- The energy storage system also plays a crucial role in maintaining the off-grid microgrid's voltage and frequency. More storage capacity in the energy storage system results in a minor power outage and a diesel generator's fuel cost. The energy storage system should be fully charged to preserve the energy to be used when the power generated from the RES is not available.

- When the energy storage is already fully charged, or the state of charge reaches the maximum level, the power generation in the off-grid microgrid is needed to be controlled. The proposed microgrid controller in this study is able to control the power generated from PV and micro-hydro to meet the total demand in the meantime, also maintaining to be fully charged.

- In the scenario that the power generated from PV and micro-hydro is not enough for the total demand, the proposed microgrid controller manages to maintain the microgrid by turning on the diesel generator to help support the power balance in the microgrid. However, the increase in operating cost due to the fuel cost needed to be considered.

- Load shedding function in the proposed microgrid controller can be used to manage the demand by curtailing the low priority loads from the microgrid to maintain the microgrid operation if the power generation very low compared to the demand. However, the disadvantages resulting from the outage of the curtailed loads are needed to be carefully considered.

6.2 Future work

The limit of the rule-based microgrid controller is that it is sometimes not flexible in the general case of the microgrid. The algorithm and program are needed to be adjusted and tested by the experts. When the microgrid is connected with new components such as renewable energy resources, the program and algorithm need to be updated. Other techniques to control the off-grid microgrid, such as Artificial Intelligence and Machine Learning Techniques, are required to be studied and implemented. Furthermore, the other microgrid control scheme that can provide a flexible 'plug-and-play' control scheme to the microgrid, such as a multi-agent system (MAS), is needed to study and implement off-grid microgrid as well.

Acknowledgements

I would like to acknowledge and thank the following people who have supported me, not only during the project but throughout my master degree.

Firstly, I would like to express my sincere gratitude to my supervisor, Professor Mark Sumner, for his unwavering support, guidance, ideas, and feedback throughout this project.

I would also like to special thank Provincial Electricity Authority (PEA), Thailand, for supporting a full scholarship. Also, thank PEA colleagues who support the valuable information and essential data for this project.

Finally, I would like to thank all my friends and family, especially my wife. You have all encouraged and believed in me. You have all helped me to focus on what has been a hugely rewarding and enriching process.

Conflict of interest

The authors declare no conflict of interest.

Author details

Tapparit Bangtit
University of Nottingham, Nottingham, United Kingdom

*Address all correspondence to: tapparitb@gmail.com

References

[1] J. Yaungket and T. Tezuka, "A survey of remote household energy use in rural Thailand," Energy Procedia, vol. 34, no. May 2019, pp. 64–72, 2013, doi: 10.1016/j.egypro.2013.06.734.

[2] A. A. Zaidi and F. Kupzog, "Microgrid automation - A self-configuring approach," IEEE INMIC 2008 12th IEEE Int. Multitopic Conf. - Conf. Proc., pp. 565–570, 2008, doi: 10.1109/INMIC.2008.4777802.

[3] A. Parisio, E. Rikos, and L. Glielmo, "A model predictive control approach to microgrid operation optimization," IEEE Trans. Control Syst. Technol., vol. 22, no. 5, pp. 1813–1827, 2014, doi: 10.1109/TCST.2013.2295737.

[4] S. Parhizi, H. Lotfi, A. Khodaei, and S. Bahramirad, "State of the art in research on microgrids: A review," IEEE Access, vol. 3, pp. 890–925, 2015, doi: 10.1109/ACCESS.2015.2443119.

[5] A. M. Bouzid, J. M. Guerrero, A. Cheriti, M. Bouhamida, P. Sicard, and M. Benghanem, "A survey on control of electric power distributed generation systems for microgrid applications," Renew. Sustain. Energy Rev., vol. 44, pp. 751–766, 2015, doi: 10.1016/j.rser.2015.01.016.

[6] D. Palit and A. Chaurey, "Off-grid rural electrification experiences from South Asia: Status and best practices," Energy Sustain. Dev., vol. 15, no. 3, pp. 266–276, 2011, doi: 10.1016/j.esd.2011.07.004.

[7] S. Ximena, C. Quintero, and J. D. Mar, "Planning for Operation of MicroGrids Considering Small Hydro Power Plants : Transient Stability," Ieee, pp. 3–8, 2013.

[8] M. E. Khodayar, "Rural electrification and expansion planning of off-grid microgrids," Electr. J., vol. 30, no. 4, pp. 68–74, 2017, doi: 10.1016/j.tej.2017.04.004.

[9] "Off Grid / Island Systems and Hybrid System - Intech Clean Energy, UK." https://www.intechcleanenergy.co.uk/off-grid-hybrid/ (accessed Jul. 26, 2020).

[10] G. Pepermans, J. Driesen, D. Haeseldonckx, R. Belmans, and W. D'haeseleer, "Distributed generation: Definition, benefits and issues," Energy Policy, vol. 33, no. 6, pp. 787–798, 2005, doi: 10.1016/j.enpol.2003.10.004.

[11] W. El-Khattam and M. M. A. Salama, "Distributed generation technologies, definitions and benefits," Electr. Power Syst. Res., vol. 71, no. 2, pp. 119–128, 2004, doi: 10.1016/j.epsr.2004.01.006.

[12] T. Ackermann, G. Andersson, and L. Söder, "Distributed generation: A definition," Electr. Power Syst. Res., vol. 57, no. 3, pp. 195–204, 2001, doi: 10.1016/S0378-7796(01)00101-8.

[13] H. Musa, "A Review of Distributed Generation Resource Types and Their Mathematical Models for Power Flow Analysis," Int. J. Sci. Technol. Soc., vol. 3, no. 4, p. 204, 2015, doi: 10.11648/j.ijsts.20150304.21.

[14] M. F. Akorede, H. Hizam, and E. Pouresmaeil, "Distributed energy resources and benefits to the environment," Renew. Sustain. Energy Rev., vol. 14, no. 2, pp. 724–734, 2010, doi: 10.1016/j.rser.2009.10.025.

[15] K. N. Nwaigwe, P. Mutabilwa, and E. Dintwa, "An overview of solar power (PV systems) integration into electricity grids," Mater. Sci. Energy Technol., vol. 2, no. 3, pp. 629–633, 2019, doi: 10.1016/j.mset.2019.07.002.

[16] B. Belcher, B. J. Petry, T. Davis, and K. Hatipoglu, "The Effects of Major

Solar Integration on a 21-Bus System," Technol. Rev. PSAT simulations, Conf. Proc. - IEEE SOUTHEASTCON, 2017, [Online]. Available: https://doi.org/10.1109/%0ASECON.2017.7925361.

[17] E. Mulenga, "Impacts of integrating solar PV power to an existing grid Case Studies of Mölndal and Orust energy distribution (10/0.4 kV and 130/10 kV) grids," pp. 1–142, 2015, [Online]. Available: http://studentarbeten.chalmers.se/publication/218826-impacts-of-integrating-solar-pv-power-to-an-existing-grid-case-studies-of-molndal-and-orust-energy-d.

[18] N. H. Saad, A. A. El-Sattar, and A. E. A. M. Mansour, "Improved particle swarm optimization for photovoltaic system connected to the grid with low voltage ride through capability," Renew. Energy, vol. 85, pp. 181–194, 2016, doi: 10.1016/j.renene.2015.06.029.

[19] R. K. Saket, "Design, development and reliability evaluation of micro hydro power generation system based on municipal waste water," 2008 IEEE Electr. Power Energy Conf. - Energy Innov., 2008, doi: 10.1109/EPC.2008.4763355.

[20] J. L. Márquez, M. G. Molina, and J. M. Pacas, "Dynamic modeling, simulation and control design of an advanced micro-hydro power plant for distributed generation applications," Int. J. Hydrogen Energy, vol. 35, no. 11, pp. 5772–5777, 2010, doi: 10.1016/j.ijhydene.2010.02.100.

[21] O. Paish, "Small hydro power : technology and current status," vol. 6, pp. 537–556, 2002.

[22] B. Guo et al., "Variable speed micro-hydro power generation system Review and Experimental results," Symp. Genie Electr., 2018.

[23] V. Salas, W. Suponthana, and R. A. Salas, "Overview of the off-grid photovoltaic diesel batteries systems with AC loads," Appl. Energy, vol. 157, pp. 195–216, 2015, doi: 10.1016/j.apenergy.2015.07.073.

[24] A. Elmitwally and M. Rashed, "Flexible operation strategy for an isolated PV-diesel microgrid without energy storage," IEEE Trans. Energy Convers., vol. 26, no. 1, pp. 235–244, 2011, doi: 10.1109/TEC.2010.2082090.

[25] M. U. Hassan, M. Humayun, R. Ullah, B. Liu, and Z. Fang, "Control strategy of hybrid energy storage system in diesel generator based isolated AC micro-grids," J. Electr. Syst. Inf. Technol., vol. 5, no. 3, pp. 964–976, 2018, doi: 10.1016/j.jesit.2016.12.002.

[26] H. Chen, T. N. Cong, W. Yang, C. Tan, Y. Li, and Y. Ding, "Progress in electrical energy storage system: A critical review," Prog. Nat. Sci., vol. 19, no. 3, pp. 291–312, 2009, doi: 10.1016/j.pnsc.2008.07.014.

[27] O. Ellabban, H. Abu-Rub, and F. Blaabjerg, "Renewable energy resources: Current status, future prospects and their enabling technology," Renew. Sustain. Energy Rev., vol. 39, pp. 748–764, 2014, doi: 10.1016/j.rser.2014.07.113.

[28] M. C. Argyrou, P. Christodoulides, and S. A. Kalogirou, "Energy storage for electricity generation and related processes: Technologies appraisal and grid scale applications," Renew. Sustain. Energy Rev., vol. 94, no. November 2017, pp. 804–821, 2018, doi: 10.1016/j.rser.2018.06.044.

[29] X. Wu, S. Feng, and P. Jiang, "Distributed Coordination Load Shedding of Islanded Microgrids Based on Sub-Gradient Algorithm," IEEE Access, vol. 5, pp. 27879–27886, 2017, doi: 10.1109/ACCESS.2017.2725901.

[30] M. Giroletti, M. Farina, and R. Scattolini, "A hybrid frequency/power

based method for industrial load shedding," Int. J. Electr. Power Energy Syst., vol. 35, no. 1, pp. 194–200, 2012, doi: 10.1016/j.ijepes.2011.10.013.

[31] P. Monshizadeh, C. De Persis, N. Monshizadeh, and A. Van Der Schaft, "A communication-free master-slave microgrid with power sharing," Proc. Am. Control Conf., vol. 2016-July, pp. 3564–3569, 2016, doi: 10.1109/ACC.2016.7525466.

[32] C. A. Hernandez-Aramburo, T. C. Green, and N. Mugniot, "Fuel consumption minimization of a microgrid," IEEE Trans. Ind. Appl., vol. 41, no. 3, pp. 673–681, 2005, doi: 10.1109/TIA.2005.847277.

[33] V. Prema, S. Datta, and K. Uma Rao, "An effective dispatch strategy for hybrid power management," IEEE Reg. 10 Annu. Int. Conf. Proceedings/TENCON, vol. 2016-Janua, 2016, doi: 10.1109/TENCON.2015.7372846.

[34] T. A. Jumani, M. W. Mustafa, A. S. Alghamdi, M. M. Rasid, A. Alamgir, and A. B. Awan, "Computational Intelligence-Based Optimization Methods for Power Quality and Dynamic Response Enhancement of ac Microgrids," IEEE Access, vol. 8, pp. 75986–76001, 2020, doi: 10.1109/ACCESS.2020.2989133.

[35] T. Kasirawat, P. Boonsiri, and T. Saksornchai, "PEA microgrid design for coexistence with local community and environment: Case study at Khun pae village Thailand," 2017 IEEE Innov. Smart Grid Technol. - Asia Smart Grid Smart Community, ISGT-Asia 2017, pp. 1–5, 2018, doi: 10.1109/ISGT-Asia.2017.8378343.

Section 2

A New Energy Management Method for Energy Storage Systems in Micro-Grids

Chapter 2

A Novel Energy Management Control Technique for PV-Battery System in DC Microgrids

Hadis Hajebrahimi, Sajjad Makhdoomi Kaviri, Suzan Eren and Alireza Bakhshai

Abstract

This paper presents a new energy management control technique for PV-Battery system used in DC microgrids. The proposed control technique is performed based on a droop control algorithm that maintains DC-bus voltage in a desirable and required range adaptively. Tightly Regulating the bus voltage In the islanded mode of operation is very challenging. However, the proposed control method by introducing a nonlinear droop profile with four adaptive parameters shows its superiority. Adaptive parameters determined by the non-linear optimal algorithms. Tightly regulating the DC bus voltage during extensive changes in demand loads/sources within a DC Micro Grid is the responsibility of the adaptive parameters. Stability of the proposed method in the whole system for a very broad range of operating conditions are proved. Simulation results along with the experimental results verify the feasibility of the proposed approach while demonstrate its superior performance compared to the conventional control method.

Keywords: Adaptive Droop Control, PV-Battery Systems, DC Micro-grid Systems, Nonlinear Optimization, Renewable energy

1. Introduction

The main interfaces utilized in microgrids (MGs) for Renewable Energy Sources (RESs) and Energy Storage Systems (ESSs) are power electronic converters [1, 2]. MGs' reliable operation depends on these power electronic converters [3–6]. Specifically, power converters' control systems which introduced recently, can make MG systems more reliable for future mainstream power generation.

The AC power system are more common due to its advantages for long distance power transmission (i.e., transformers can be used to adjust optimal voltage levels). Thus, the development of micro-grid technology has been mostly confined to AC micro-grids [4, 5]. However, as the number of local micro-grids increase the need for long distance power transmission will significantly decline. Therefore, local DC micro-grids can potentially be more efficient due to the fact that many loads require DC power (e.g., electronic loads, LED lighting systems, electric vehicle charging, etc.). In addition, most renewable energy systems and energy storage systems are either inherently DC (batteries, solar panels) or incorporate a DC stage (inverter connected wind turbines) [6]. Since the number of DC sources and loads will inevitably increase in the future, the need for a DC MGs will also increase. The reason is that the need for PV cells and ESSs is increasing. The advantages of the DC MGs over the AC MGs are summarized as follows [7, 8]:

1. DC MGs are more efficient and the natural fits for DC sources and loads.

2. Most of the existing challenges in AC MGs are due to reactive power flow, power quality, and frequency regulation, which do not exist in the DC MGs.

3. In terms of transmission efficiency (loss associated with the reactive current is eliminated) and power supply reliability (more reliability overall because of fewer components), DC MGs have a superior performance in comparison to AC MGs [8].

Thus, DC microgrids seem to be a natural framework for future power delivery [9].

In DC MGs, regulating the common DC-bus voltage is the main control task. To gain this goal, two different methods have been proposed in the literature [10–14], autonomous droop-based control schemes and non-autonomous centralized controllers (that are based on communication links). Droop-based methods are widely used due to their simplicity and reliability (centralized methods can not be relied upon due to their dependency on the communication links) [15–31]. However, the existing and conventional droop-based methods cannot propose an optimal performance in terms of reliability and efficiency exclusively in the islanded mode of operation. In addition, existing droop controllers usually result in large voltage variations at the DC bus during transients in the islanded mode. Large voltage fluctuations may jeopardize the system reliability and damage power electronic converters in MGs [3, 4]. However, the proper performance of linear or conventional droop controllers with a constant droop gain is reduced in light and full load conditions. This is due to a trade-off between voltage deviation and current sharing. In terms of current sharing accuracy and stabilization of Constant Power Loads (CPLs), high droop gains are preferable [11]. In term of voltage regulation low droop gains are desirable. Due to this trade-off between current sharing and voltage deviation, constant droop gains can not guarantee high performance, and it would be a big challenge. **Figure 1** shows this trade-off in the conventional droop control in a simplified DC MG's with two DC sources. It shows that constant droop gain can not be a precise and comprehensive method in both light and full load conditions [11].

To address this issue, different methods have been proposed in the literature to maintain a high level of reliability [5, 6]. A fuzzy logic strategy for energy balancing is proposed in [12]. This method is based on utilizing the virtual resistances which are calculated based on the *SoC* of each energy storage unit using fuzzy logic algorithms in the droop profile to address the reliability and energy management issues. The prominent disadvantage of this method is the high amount of voltage deviations and circulating currents. Another energy management method that is recently presented in [13] is the mode-adaptive droop controller. Although this method can alleviate the energy management issue, it is not very sensitive to the small values of voltage differences in the voltage sensors. An adaptive droop control strategy is presented in [14], which provides load sharing and circulating current minimization. This method is based on the interface resistance between the converter terminal and the DC bus. Thus, the resistance values should be known in advance or calculated, which is a relatively complex process. A distributed adaptive-droop control with battery management capability is presented in [15].

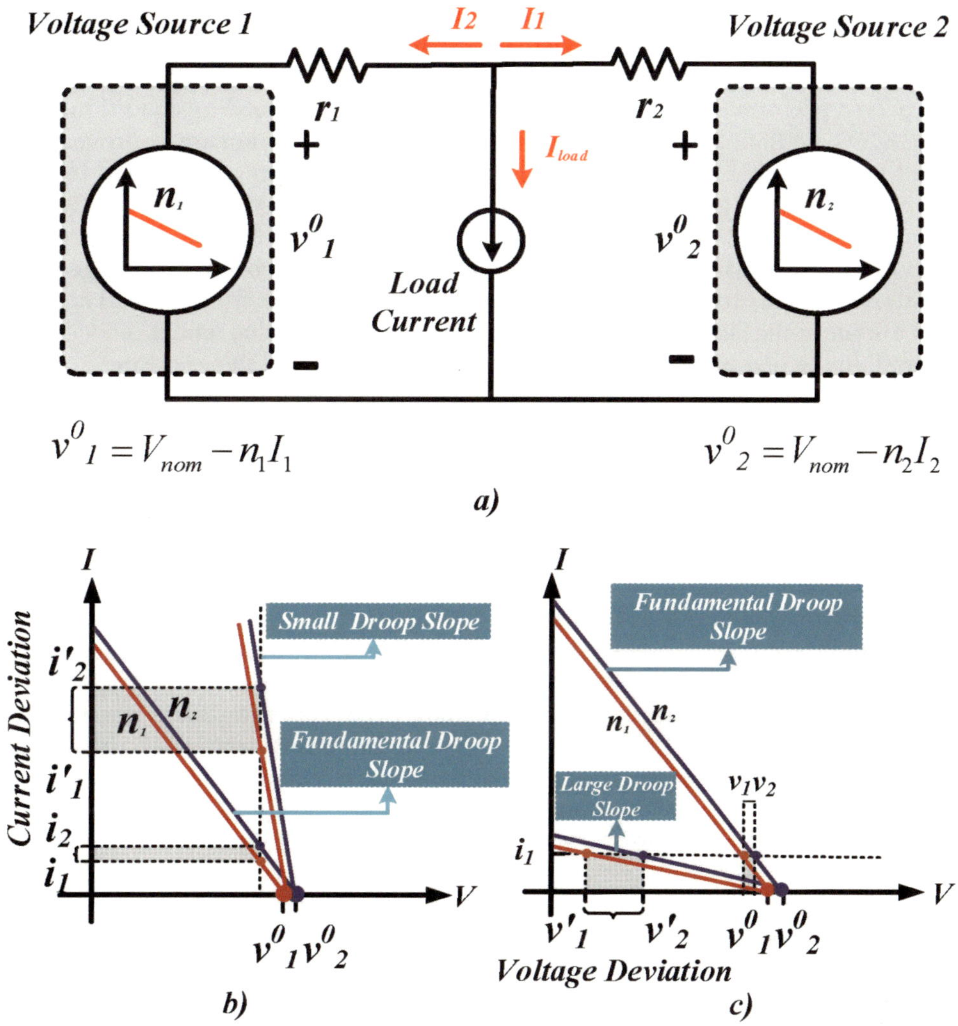

Figure 1.
(a) A simple DC MG with two DGs, (b) the effect of small drop causes voltage deviation with current splitting error at full load, and (c) the effect of the large droop gains in voltage deviation with current sharing error under light load conditions, respectively.

In this paper investigate a new double-layer hierarchical control strategy in MG system. The primary control layer is based on an adaptive voltage-droop profile that balances the common bus voltage while maintaining the *SOCs* of batteries close to each other [15]. The second layer uses a low bandwidth communication link to perform supervisory control tasks. The communication link is used to collect the required data to calculate the adaptive virtual resistances (VRs) along with the transit criteria for changing unit-level operating modes. In comparison to the decentralized controller methods, communication-based controllers need an infrastructure along with an increase in cost and unreliability.

Three nonlinear droop control methods are proposed in [17–19]. In [17], it is claimed that the effect of sensor calibration errors and cable resistances is minimized, while it improves voltage regulation and load sharing compared to linear droop control techniques. This method adaptively adjusts the droop gains based on three novel high-order polynomial droop equations. The three methods are proposed in this research are High-Droop Gain (HDG), Polynomial Droop Curve (PDC), and Polynomial Droop Curve With Voltage Compensation (PDCVC), which are shown in **Figure 2**. However, the chosen droop gain value in the HDG method exhibits poor current sharing near the operating point. On the other hand, the PDC method could improve load sharing near the operating point, but shows an unexpected performance in voltage regulation under heavy-loading condition. In the PDCVC method, both current sharing and voltage regulation are improved except near no-load conditions. Two important factors were not considered in this paper, the impact of output impedance on the stability of the whole DC MG and the impact of droop control on the efficiency of system. Later in [18], a split droop controller is investigated, which droop gain can be calculated based on voltage error measurement. This strategy calculates several different slopes for each load conditions to reduce the current sharing and voltage deviation errors, which shows this task well implemented. However, differences between droop slopes create switching modes, which will be abrupt. The sudden changes in output resistance of the power converters may lead to undesired transients and oscillations. In [19], the performance of the different second order droop expressions namely parabola and ellipse droop as nonlinear droop control methods are evaluated and compared with the linear droop. The parabola and ellipse droop equations showed a good performance in both load sharing and voltage regulation but a weak performance with negative current [16]. A nonlinear programmable droop method for a low-voltage DC MG is proposed in [20] considering the effects of line resistances and the state of charge of the batteries in the power- sharing process. This method has been able to reduce the impacts of sensor and cable errors at heavy load. It also reduces the voltage variation while load sharing is still guaranteed. However this method does not have a good performance, when the converter output current is negative. Moreover, under light-loading conditions, the values of droop gain lead to creating poor load sharing [20].

In this paper a novel control system has been proposed to improve the reliability and efficiency of DC micro-grids operating in the islanded mode. **Figure 3** shows the block diagram of the DC MG considered in this paper. Two energy storage systems along with two solar energy harvesting systems are connected to the DC bus to feed 4 kW DC loads. The focus of this paper is on small scale DC micro-grids for residential applications, which means the line impedance can be negligible [14, 19]. The main contribution to this article is that the proposed control system allows RESs to operate at their maximum power point whenever it's possible. The main advantage of the proposed control system is that the MG can optimally utilize energy storage system in conjunction with renewable energy systems. Additionally, the proposed control system can maintain tight regulation of the DC bus voltage within a predefined range. Unlike the aforementioned nonlinear methods, the presented method has a precise performance in negative current (when the output

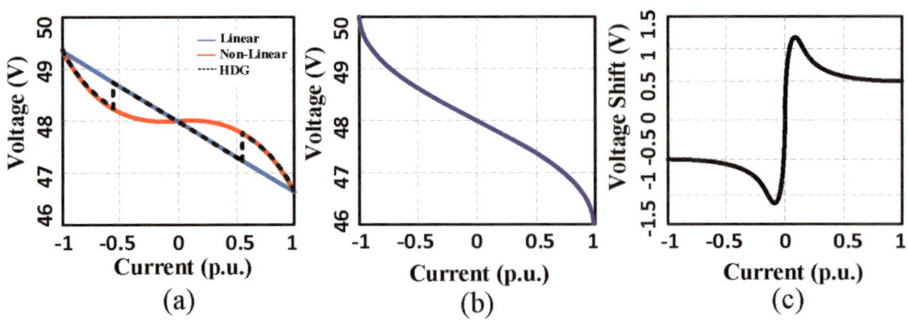

Figure 2.
(a) Droop characteristics for HDG method, (b) droop characteristics for PDC method, and (c) Δv curve for PDCVC method.

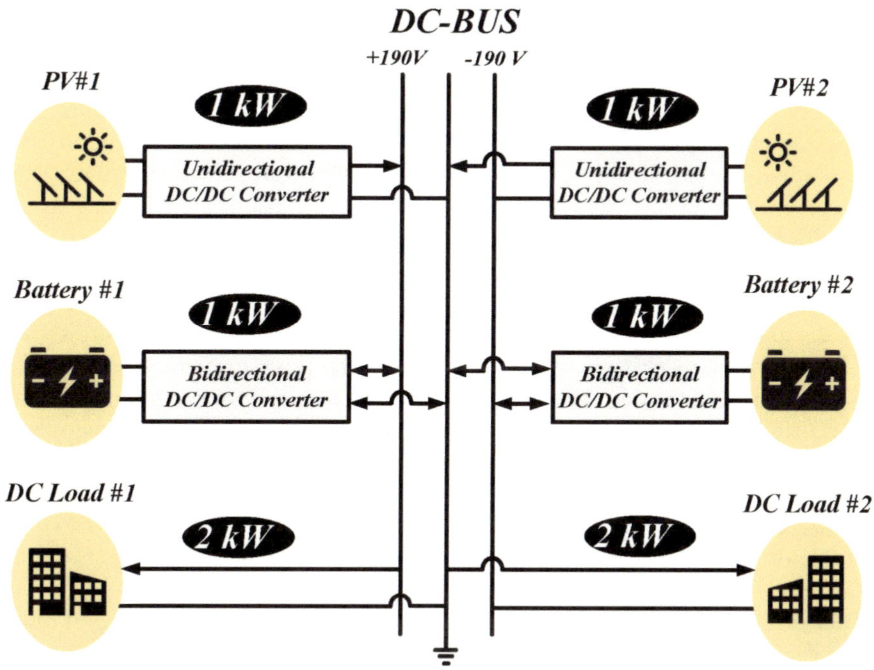

Figure 3.
DC microgrid configuration (implemented system).

power is absorbed by the battery) which is similar to the positive current (when the battery converter injects power to the MG). The calculations of the droop gain in the proposed method are independent of the measured output current. Therefore, this method immunizes the proposed control system from current ripples and sensor errors in comparison to the methods with high order polynomial droop methods. This method has an ability to change the droop gain smoothly from light load to full load conditions, without switching modes resulting in no abrupt in the control system. Voltage deviation and current sharing improved in this method in comparison to other nonlinear and conventional droop control, and it will be shown and explain in the following section. This method has been implemented and analyzed in the Micro-Grid systems due to this fact, the most difficult tasks and challenges have occurred in the MG system, not in the grid- connected system. In the grid-connected system the bus voltage is constant with less complexity and less non-ideal elements or parameters. Various power flow scenarios have been analyzed to verify the performance of the MG.

This paper is organized as follows. The proposed control system and various operating scenarios are described in Section 2, respectively. Moreover, two subsections called the definition and ranges of the adaptive parameter are added in Section 2. In Section 3, the optimal selection of the the adaptive parameters is presented. The stability analysis performed in Section 4. Finally the simulation and experimental results are presented in Section 5.

2. Proposed control system

In order to describe the performance of the proposed control system in all operation modes, a scenario will be discussed. Energy devices and systems in which the solar system and battery are connected to the grid/load through power converters have four operating modes which are listed as follows:

1. Grid-connected mode: When the system is connected to the grid, in order to transmit the maximum power, the power converter connected to the solar cells performs maximum power point tracking (MPPT). During this mode, the battery's state of charge should be controlled so that when the system is disconnected from the grid and turns into an island system, the battery has enough energy to supply the island system. Therefore, in accordance with the battery's state of charge, the battery is charged or discharged, which are two modes of operation.

2. Island mode: In this mode, the power converter connected to the solar cells still performs MPPT. During this mode of operation, the DC-bus voltage should be kept constant (within the range of 10%) by the power converter; whereas, in grid-connected mode, the voltage of DC-bus is relatively constant for various conditions. Therefore, to achieve power-sharing and load voltage regulation, the battery connected power converter is controlled with the droop control method. For instance, if the load voltage decreases, battery is discharged to transmit more power; and when the load voltage increases, power is transmitted back to the battery and charges it.

3. Special case of island mode: If the battery is fully charged/discharged during the island mode, the power-sharing and output voltage regulation are achieved by the solar connected power system. Hence the power system evades MPPT and is controlled with the droop control method.

In the islanded mode of operation, one of the main problems with the existing droop-based controller is that the MG components are not utilized in an efficient manner [30, 31]. For example, in several scenarios, to keep the stability of the system, renewable energy sources must deviate from their maximum PowerPoint [30, 31]. The main contribution of this method shows the presented control system allows RESs to operate at their maximum power point whenever it can be possible. **Figure 4** shows the general block diagram of the proposed control systems for the solar energy harvesting system and the energy storage system (ESS) in the DC MG. The solar control system consists of one MPPT control and one adaptive droop control, which are separated from each other by a state modifier. The structure of the solar control system will be discussed in the next paragraph. After that, the battery control system will be discussed in details as the main proposed control method. The battery control system consists of a supervisory adaptive droop control to manage energy through the DC MG system. The optimal utilization is performed through incorporating the adaptive charge/discharge terms to the control system of the ESS. These terms implement a nonlinear adaptive droop profile for the ESS in order to perform a tighter voltage regulation. Moreover, due to working in a high level of power (More than 100 W), all the analysis are preferred in continues conduction modes, because working in discontinuous conduction mode in high-level power comes with high peak current and power loss, which it is not desired.

According to **Figure 5**, the schematic of the solar controller along with its boost converter is illustrated. The outer control loop of the solar energy harvesting system consists of two blocks: the MPPT controller that ensures maximum power is extracted from the solar module, and the adaptive DC-bus voltage controller. The switching between these two control algorithms is preformed by the state modifier block as follows:

$$\begin{cases} if \;\; V_L \leq \; v_{DC} \leq \; V_H \Rightarrow State := MPPT \;\; Mode \\ Otherwise \Rightarrow State := Droop \;\; Mode \end{cases} \tag{1}$$

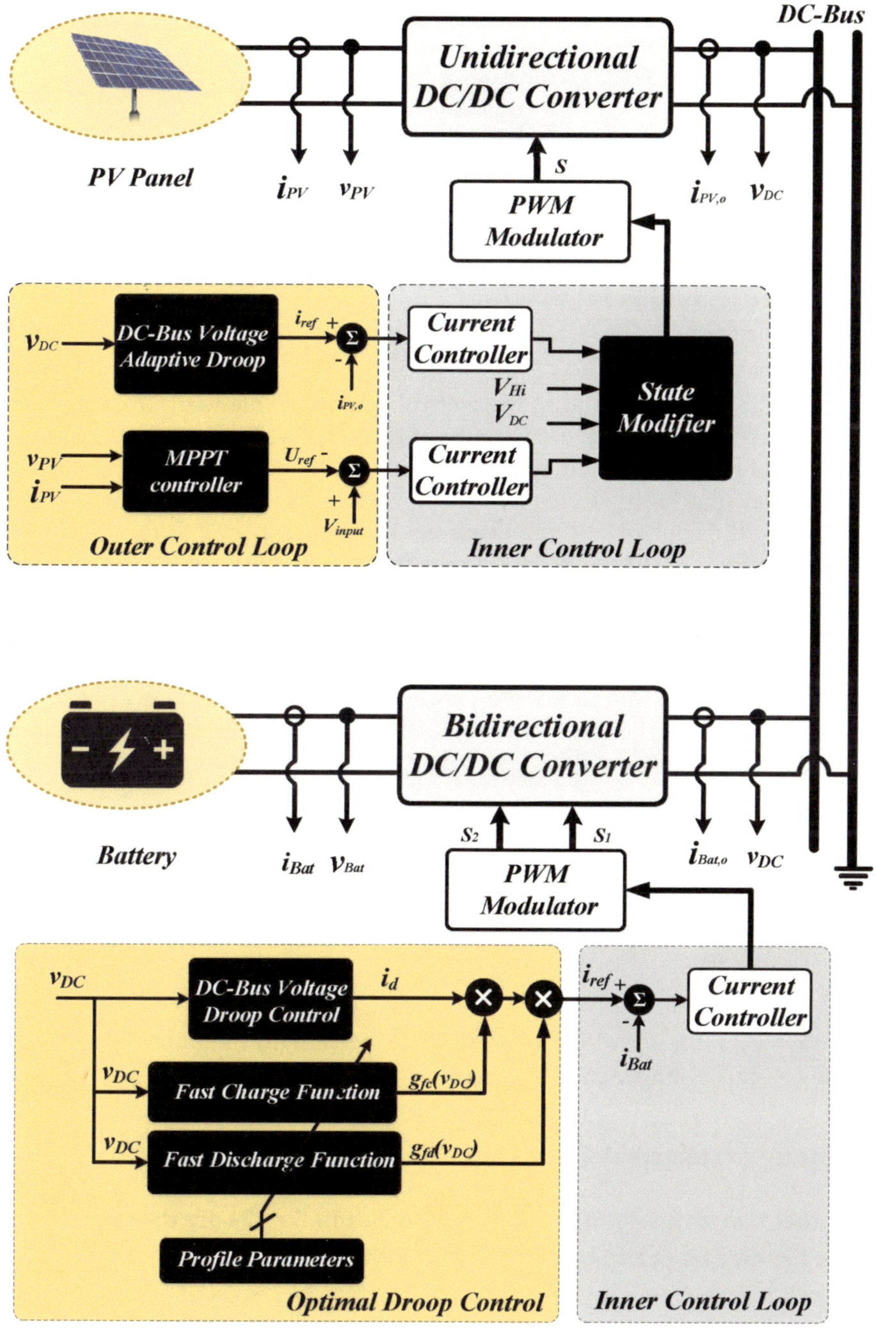

Figure 4.
Block diagram of the proposed control scheme for PV units and energy storage systems in DC MG.

where V_L and V_H are defined based on the MG requirements and safety standards [29]. This structure provides flexibility to operate at the MPPT as much as possible.

The block diagram of the ESS is discussed according to the **Figure 4**. In this figure, the presented control method offers four modes of operations: charging mode, discharging mode, fast charging mode, and fast discharging mode while the conventional droop control method offers only, charging mode and discharging

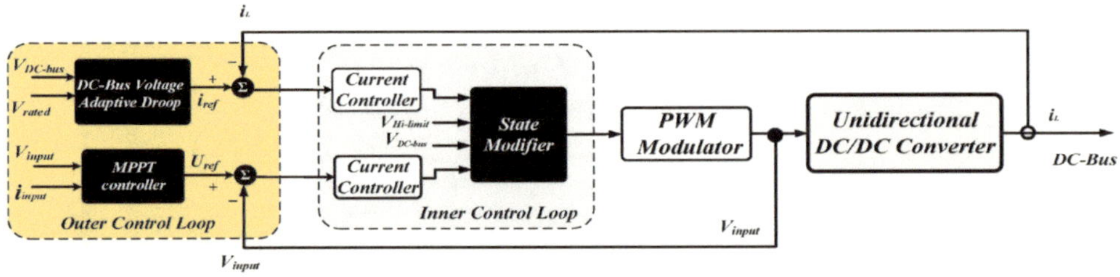

Figure 5.
The block diagram of control system in solar converter.

mode. Using the two extra modes, the control system is able to offer voltage regulation, while improving the reliability of the system. The outer loop produces a reference value for the inner current control loop. The outer loop is optimized droop controller that incorporates two extra modes (i.e., fast charge, and fast discharge) into the control system. These modes are incorporated through two nonlinear functions, $g_{fc}(v), g_{fd}(v)$ (fc) stands for fast charge and (fd) is short for "fast discharge"). These functions effectively determine the speed of charge and discharge of the battery based on the DC-bus voltage.

Control systems based on optimization methods can be the proper solutions in determining the critical and unknown parameters of the MG systems [28, 29]. According to **Figure 3**, the optimal droop controller produces the reference value for the output current of the ESS. This current is given by:

$$i_{ref} = i_d(v_{DC})g_{fc}(v_{DC})g_{fd}(v_{DC}) \tag{2}$$

The reference current includes three terms. The first term is produced by the DC-bus voltage droop controller given by:

$$i_d = -k_D \times (V_{nom} - v_{DC}) \tag{3}$$

where k_D, V_{nom}, v_{DC}, and i_{ref} are the droop coefficient, the nominal voltage value, the DC-bus voltage, and the reference current respectively.

2.1 Adaptive parameters definition

The other two terms, $g_{fc}(v), g_{fd}(v)$, represented in Eq. (2) are the adaptive functions for the fast charging and fast discharging modes, given by (when $V_{min} \le v_{DC} \le V_{max}$):

$$g_{fc}(v_{DC}) = \exp\left(\frac{(v_{DC} - V_{fc})}{\alpha_{fc}} u(v_{DC} - V_{fc})\right) \tag{4}$$

$$g_{fd}(v_{DC}) = \exp\left(\frac{(V_{fd} - v_{DC})}{\alpha_{fd}} u(V_{fd} - v_{DC})\right) \tag{5}$$

where $\exp(.)$ is the exponential function, $u(.)$ is the unit step function, V_{fd} and V_{fc} both determine the start points of fast charging and fast discharging modes respectively, and α_{fc} and α_{fd} are the exponential coefficients to create fast charging and fast discharging respectively.

Figure 6 shows the way to calculate the current reference based on the DC-bus voltage for the Battery or ESS. This figure shows, the proposed controller provides two extra modes (i.e., fast charge and fast discharge), when the DC-bus voltage is lower than V_{fd} or higher than V_{fc}. When the DC-bus voltage is lower than V_{fd}, it means that the load is too heavy and the ESS needs to inject much more power to sustain the DC-bus voltage. When the DC-bus voltage is higher than V_{fc}, it means that the generation is higher than the load. Thus, the extra power is used to fast charge the battery. The adaptive droop profiles go through the fast charge/discharge modes when the step functions are equal to one. This function allows the control system to absorb and inject more power than the conventional system automatically to prevent the voltage deviation. **Figure 7** shows the performance of the proposed droop profile in current sharing and voltage deviation in a DC MG with two DC sources. Even in the worst case (when the droop gain in the two DC sources is considered equal and constant) the current sharing and voltage deviations are improved reasonably in comparison to the conventional ones in **Figure 1**.

2.2 Adaptive parameters range

In order to achieve a stable and reliable system, the limitation on the numerical domain of Eq. (2) should be considered. The aforementioned limitation is set to meet the requirements of the implemented control system. The considered limitation, which is defined according to voltage regulation range, resources power rating, and load demanded power, will be discussed in detail as follows. In the **Table 1** all the implemented components and variables are listed. The natural Logarithm domain and charge/discharge rate limitations of the implemented batteries, specified the values of α_{fd}, α_{fc}, V_{fc}, and V_{fd}. According to the simulation of the proposed control system in PSIM software and charge/discharge rate limitations of the

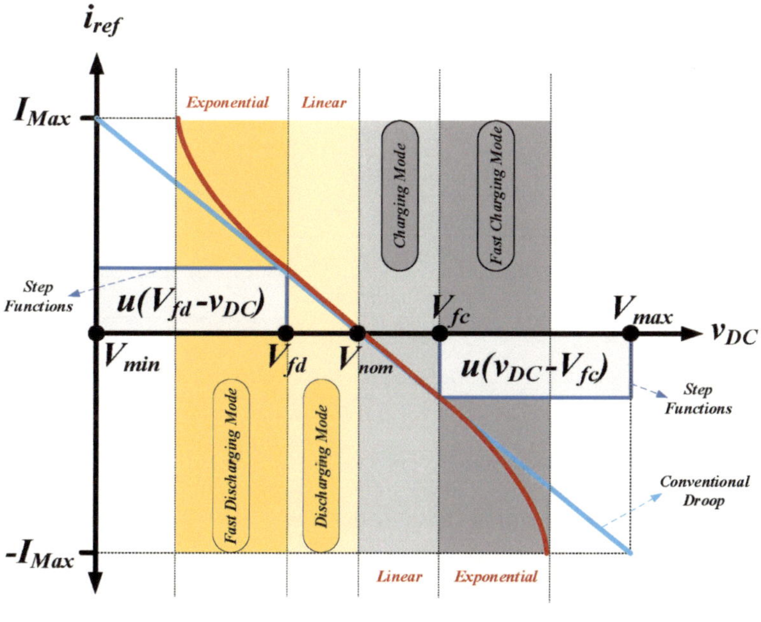

Figure 6.
The performance of the step and exponential functions in the proposed controller.

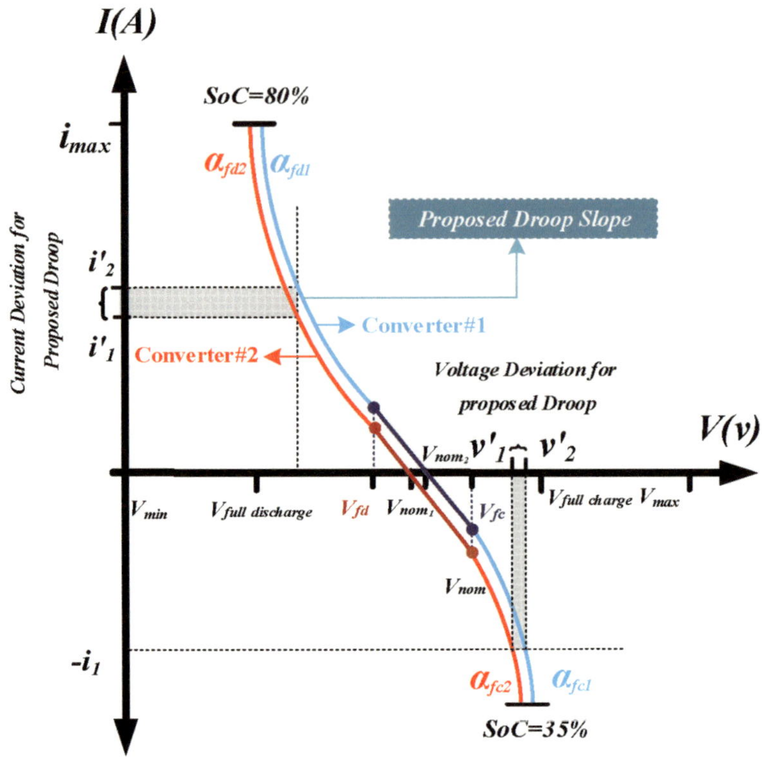

Figure 7.
The current sharing and voltage deviations in the proposed system in comparison to the conventional one in ***Figure 2****.*

Symbols	Parameters	Values
P_o	Nominal Output Power for Battery#1 converter	1 kW
P_o	Nominal Output Power for Battery#2 converter	1 kW
P_o	Nominal Output Power for PV#1 converter	1 kW
P_o	Nominal Output Power for PV#2 converter	1 kW
f_{sw}	Switching Frequency	100 kHz
V_{nom}	Nominal DC MG Voltage	190 V DC
$V_{in_{Bat}}$	Battery Bank Voltage	80 V DC
$V_{in_{PV}}$	PV Bank Voltage	80 V DC
L_1	Battery input Inductor	150μH
L_2	PV input Inductor	330μH
C_o	Output Capacitor	200μF

Table 1.
DC/DC converter specifications in simulation.

batteries, lower range of α_{fd} and α_{fc} have to be specified to ensure the reliable operation of the system in all operating conditions as follow:

$$\alpha_{fc} \geq 2.9, \tag{6}$$

$$\alpha_{fd} \geq 2.9, \tag{7}$$

The upper range of α_{fd} and α_{fc} are specified based on the voltage regulation range and minimum fluctuations around the rated voltage (to achieve this goal, the $V_{fc_{Min}}$ and $V_{fd_{Max}}$ must be specified to keep the minimum distance from the rated voltage). The aforementioned upper ranges are given as:

$$i_{d_{Max}} = -k_D \times (V_{nom} - v_{DC_{Max}}) \times g_{fc}(v_{DC_{Max}})\Big|_{V_{fc_{Min}}} \tag{8}$$

$$i_{d_{Max}} = -k_D \times (V_{nom} - v_{DC_{Min}}) \times g_{fd}(v_{DC_{Min}})\Big|_{V_{fd_{Max}}} \tag{9}$$

Finally the upper and lower range of α_{fd} and α_{fc} are:

$$2.9 \leq \alpha_{fc} \leq 12.8, \tag{10}$$

$$2.9 \leq \alpha_{fd} \leq 12.8, \tag{11}$$

In the implemented control system, the proposed range of the DC-bus voltage is considered to be $180 \leq v_{DC-bus} \leq 200$. Thus, the ranges for the fast charge/discharge voltages are given by:

$$V_{fd} \geq \alpha_{fd_{Min}} \times Min\left|\ln \frac{i_{ref} \times SoC(t)}{(-k_D \times (V_{nom} - v_{DC}))}\right| + v_{DC_{Min}} \tag{12}$$

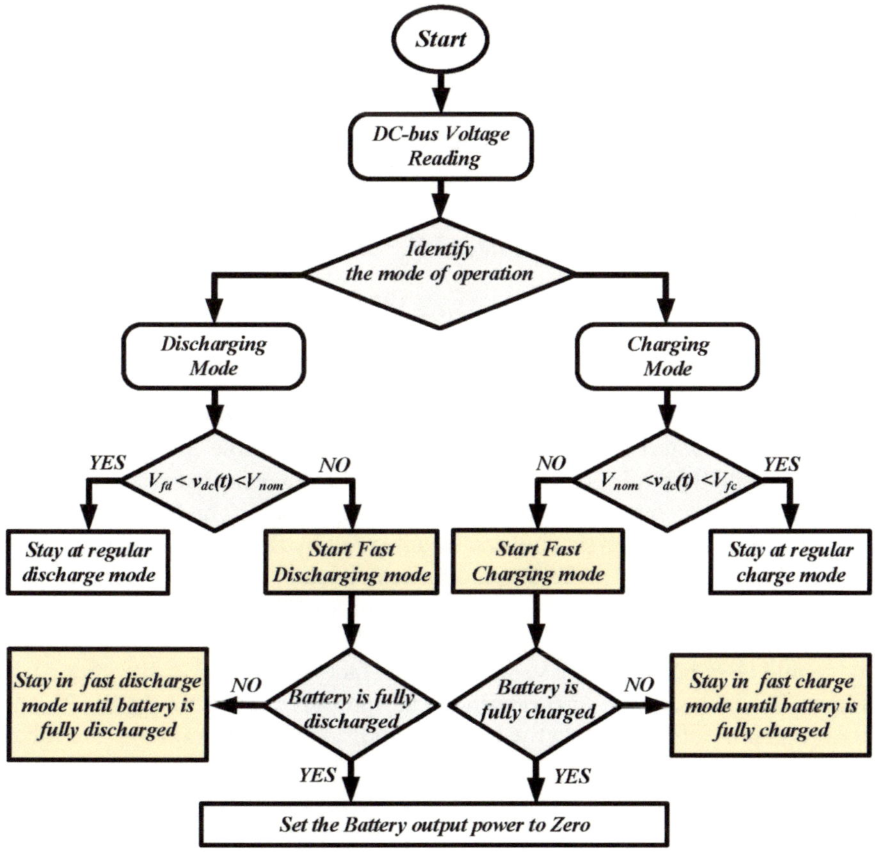

Figure 8.
The flowchart of the proposed control scheme for the ESS.

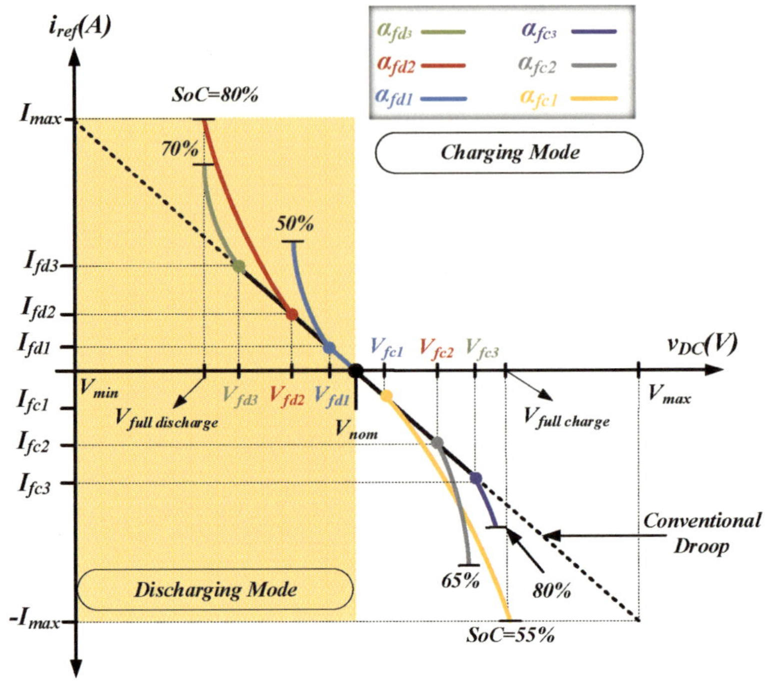

Figure 9.
Proposed adaptive droop controller with adaptive charge/discharge functions.

and

$$V_{fc} \geq v_{DC_{Max}} - \alpha_{fc_{Min}} \times Max \left| \ln \frac{-i_{ref} \times SoC(t)}{(-k_D \times (V_{nom} - v_{DC}))} \right| \tag{13}$$

Therefor, the upper and lower range of V_{fd} and V_{fc} are:

$$191 \leq v_{fc} \leq 198, \tag{14}$$

$$182 \leq v_{fd} \leq 189, \tag{15}$$

The main advantages of the profile proposed in **Figure 5** are twofold. It provides a much better dynamical result for the ESS to utilize the available energy optimally. It also provides a small margin in the range of DC-bus voltage or its variations. Thus, it gives a chance for the converters to operates more optimally. The proper voltage deviation and desired power-sharing can be achieved by this method adaptively and autonomously, without any switching modes which lead to the harsh transients. **Figure 8** shows the flowchart of the control system for ESS. This figure shows how various modes of operation get activated by the DC-bus voltage. The (*SOC*) of the battery should be considered as a very important factor in selecting different parameters. **Figure 9** depicts different profiles in the various *SOC* values. In **Figure 9** charging and discharging is determined by the battery conditions to maintain the bus voltage.

3. Optimal droop control parameters

In this section, the procedure to determine the optimal values of the parameters for the proposed optimal droop controller is described. The objective is to find appropriate values for four parameters: V_{fc}, V_{fd}, α_{fc}, and α_{fd}. The *SOC* of the ESS plays a very important role in the calculation of the adaptive parameters. An optimization algorithm to determine the precise values for the speed and depth of charge/discharge is the main idea of this method.

The proposed algorithm is based on sequential quadratic programming (SQP) [33], which is an iterative method for nonlinear optimizations. It can handle any degree of non-linearity, including non-linearity in the constraints [32]. **Figure 10** shows the block diagram of the energy storage system with the proposed control system with the all details. This diagram shows the nonlinear optimization block with its inputs and outputs. **Figure 11** shows that the SQP algorithm calculates the optimal values for the adaptive parameters. The proposed controller uses these values to generate the nonlinear functions $g_{fc}(v_{DC})$ and $g_{fd}(v_{DC})$. The following steps are implemented in the SQP algorithm:

1. The optimization variables are defined as:

$$X = \begin{bmatrix} x_1 \\ x_2 \\ x_3 \\ x_4 \end{bmatrix} = \begin{bmatrix} \Delta V_{fd} \\ \alpha_{fd} \\ \Delta V_{fc} \\ \alpha_{fc} \end{bmatrix} \tag{16}$$

where $\Delta V_{fd} = V_{fd} - V_{nom}$ and $\Delta V_{fc} = V_{fc} - V_{nom}$. As written in Eq. (16), the four optimization variables are ΔV_{fd}, ΔV_{fc}, α_{fd}, and α_{fc}. The reason for choosing these four optimization variables is that in each value of *SoC*, the control system will determine a new optimum point in the charge and discharge modes of operation. This task can not be possible without these main variables. For reliable

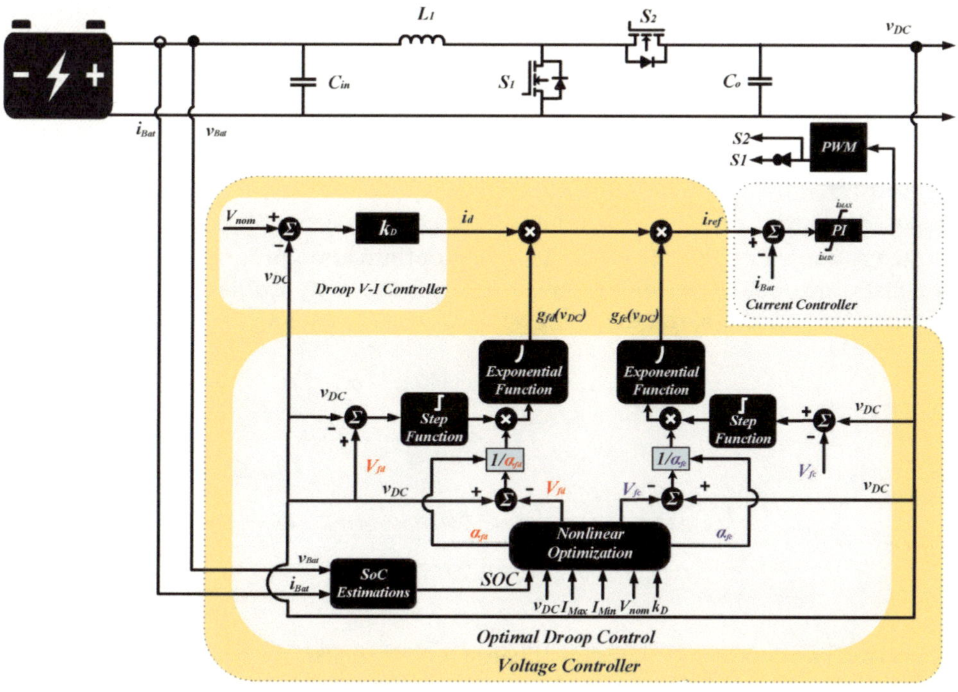

Figure 10.
The DC/DC converter for the ESS with its control system.

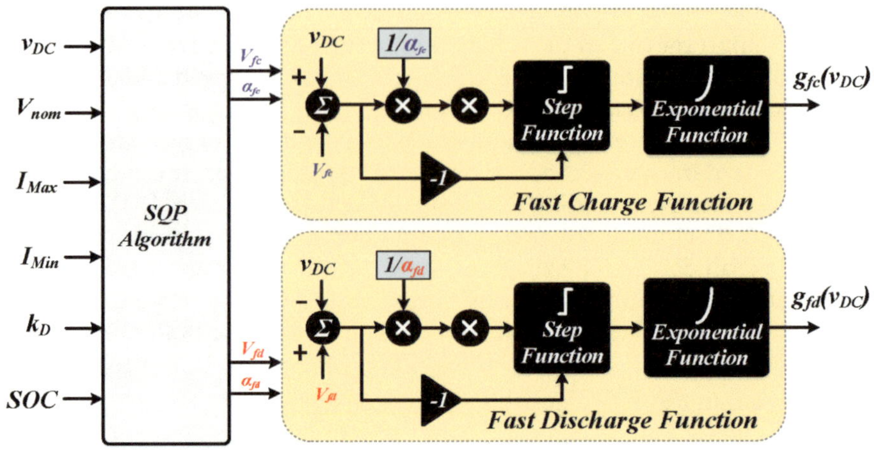

Figure 11.
Structure of charge/discharge control functions in the proposed method.

performance, three sensitive cases are proposed for system decisions in the optimization process. Two cases occur when the *SoC* value is high in discharging and charging modes, and another is when the value of *SoC* is low. For instance in discharge mode, the values of two variables (α_{fd} and ΔV_{fd}) with the highest *SoC* value will be specified by the optimization algorithm when α_{fd} has its minimum value and ΔV_{fd} has its maximum value. This decision is helpful to discharge the battery smoothly during a long time while achieving less voltage deviation.

2. The general form of the objective function is defined as:

$$SQP: \begin{cases} Minimize\ f(X) \\ Subject\ \ to: \begin{cases} H_{eq}(X) = 0 \\ H_{ineq}(X) \leq 0 \end{cases} \end{cases} \tag{17}$$

where $H_{ineq}(X)$ and $H_{eq}(X)$ are the inequality and equality constraints. In the proposed adaptive droop control, only two inequality constraints H_{fd}, and H_{fc} are defined as shown in Eqs. (18) and (19). This optimization program lacks the equality constraints, due to current or power changes in droop profile. In this particular optimization problems, the following constraints are considered:

$$H_{fd}(\Delta V_{fd}, \alpha_{fd}) = k_D \times (V_{nom} - v_{DC}(t)) \times e^{u(-v_{DC}(t) - \Delta V_{fd} + V_{nom}) \times \frac{(-v_{DC}(t) - \Delta V_{fd} + V_{nom})}{\alpha_{fd}}} - I_{max} \leq 0 \tag{18}$$

$$H_{fc}(\Delta V_{fc}, \alpha_{fc}) = I_{min} - k_D \times (V_{nom} - v_{DC}(t)) \times e^{u(-v_{DC}(t) + \Delta V_{fc} - V_{nom}) \times \frac{(-v_{DC}(t) + \Delta V_{fc} - V_{nom})}{\alpha_{fc}}} \leq 0 \tag{19}$$

Having two modes of operation (charge and discharge modes) and more than one optimization variable, the need for the cost function to optimize more than one variable simultaneously, will be increased. The cost functions for this optimization problem are defined as:

1. Discharging mode:

$$f_1(x) = \Delta V_{fd} = V_{fd} - V_{nom} \tag{20}$$

$$f_2(x) = \alpha_{fd} \tag{21}$$

2. Charging mode:

$$f_1(x) = \Delta V_{fc} = V_{fc} - V_{nom} \tag{22}$$

$$f_2(x) = \alpha_{fc} \tag{23}$$

The total cost function is defined as the weighted sum of the aforementioned functions given by:

$$f(x) = \sum_{m=1}^{2} w_m f_m(x) = w_1 f_1(x) + w_2 f_2(x) \tag{24}$$

where w_1 and w_2 are the coefficients utilized to determine the role of each function in the nonlinear optimization algorithm.

The *SOC* of the battery is also incorporated into the optimization algorithm. The multi-objective optimization function is modified to incorporate the *SOC*:

1. Case 1:

$$\begin{gathered}(Discharge\ \ mode)\\ if\ \ SOC \le 40\%\\ Maximize\ \ f(x) = w_1 \Delta V_{fd} + w_2 \alpha_{fd}\end{gathered} \tag{25}$$

2. Case 2:

$$\begin{gathered}(Discharge\ \ mode)\\ if\ \ SOC \ge 40\%\\ Minimize\ \ f(x) = w_1 \Delta V_{fd} - w_2 \alpha_{fd}\end{gathered} \tag{26}$$

3. Case 3:

$$\begin{gathered}(Charge\ \ mode)\\ for\ \ all\ \ SOC\ \ values\\ Minimize\ \ f(x) = w_1 \Delta V_{fc} - w_2 \alpha_{fc}\end{gathered} \tag{27}$$

According to (25) and (26), two operating cases have been considered for the discharge mode based on *SOC* values. In case one, the storage system needs to be discharged as slow as possible to prevent storage capacity depletion. This can be done by maximizing $f_1(x)$, and $f_2(x)$. In Case two, simulates the discharging mode when the *SOC* is more than 40%. In this case, the optimization algorithm tries to maximize $f_2(x)$ while minimizing $f_1(x)$. Also, the Eq. (27) is defined for the cases with all *SOC* values in charge mode, which means the controller system always tends to charge the ESS as fast as possible. **Figure 12** shows an exemplary optimization in case one with $SOC = 20\%$. In this figure, the total function evaluations are

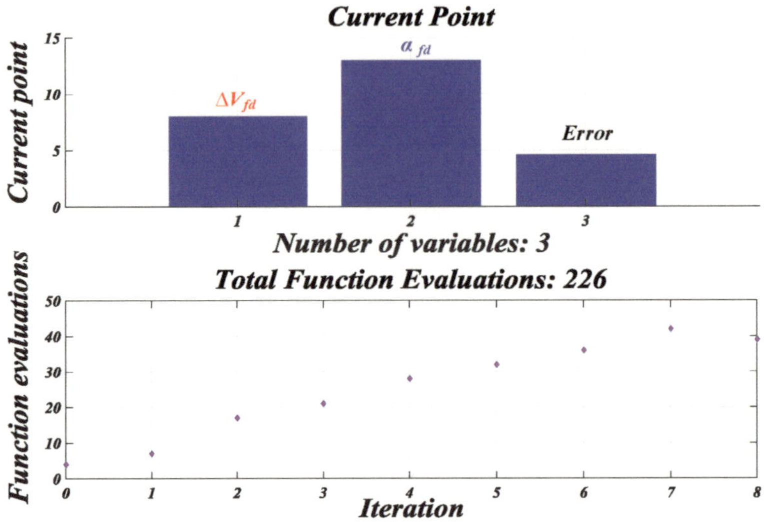

Figure 12.
An example of SQP algorithm performance in the situation of SOC = 20%.

226 which means around 226 intermediate calculations take place to reach to the next optimum solution during 8 times iterations. After 8 iterations the optimization process stops. There is an initial error-check that evaluates the objective function before any iterations take place. Three variables are defined in this figure, which are ΔV_{fc}, α_{fc}, and Error. In the case of $SOC = 20\%$ the optimum solutions are calculated for these three variables 8, 13, 5 respectively. The optimum solutions show that for this *SoC* the control system performs like the conventional droop in order to not missing the remaining power in the battery so fast.

4. Stability analysis

In this section, the stability analysis of the proposed control method is performed. The proposed optimal droop control system utilizes profiles with linear and nonlinear parts. When the system operates in the linear part (i.e., conventional droop profile), the stability analysis is straight forward. However, nonlinear methods (e.g. the Lyapunov stability theory) must be used to analyze the stability of the nonlinear part (the stability analysis has been carried out only for the islanded mode of operation) [34, 35].

Figure 13 shows the block diagram of the closed-loop control systems and the schematic diagram of the converter. According to **Figure 13**, the dynamical model of the battery is also considered in the schematic block diagram. According to **Figure 13**, the state-space model of the ESS is given by [29]:

$$\begin{cases} \dfrac{di_L}{dt} = \dfrac{1}{L} v_{C_{in}} - \dfrac{1}{L} v_{DC}.d' \\ \dfrac{dv_{DC}}{dt} = \dfrac{1}{C_o}.d'.i_L - \dfrac{1}{C_o.R_{eq}} v_{DC} \\ \dfrac{dv_{C_b}}{dt} = \dfrac{V_{oc} - v_{Cb} - v_{C_{in}}}{C_b.R_{sb}} - \dfrac{v_{C_b}}{C_b.R_b} \\ \dfrac{dv_{C_{in}}}{dt} = \dfrac{V_{oc} - v_{Cb} - v_{C_{in}}}{C_{in}.R_{sb}} - \dfrac{i_L}{C_{in}} \end{cases} \tag{28}$$

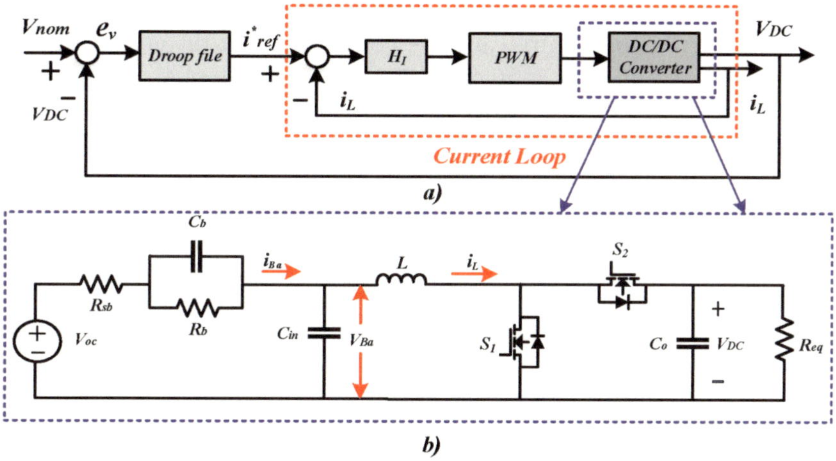

Figure 13.
(a) Closed-Loop block diagram of the battery controller. (b) DC-DC converter schematic.

where, C_{in}, C_b, C_o, R_{eq}, R_{sb}, and L are the input capacitor of the DC/DC converter, the equivalent series capacitor in the battery model, the DC/DC converter output capacitor of battery, the MG side equivalent load, and the DC/DC converter input inductor. v_{DC}, $v_{C_{in}}$, v_{C_b}, i_L, and i_{Ba} are the DC bus voltage, the input voltage of the DC/DC converter, the voltage across C_b, the inductor current and the input current of the battery respectively. Also, $d' = 1 - d$, where d is duty cycle.

According to (28), the system dynamics include fast changing current, i_L, and slow changing voltages, v_{DC}, $v_{C_{in}}$, v_{C_b}. Due to different rates of changes, the cascade control system shown in **Figure 13** includes the fast inner current loop and the slow outer voltage loop. The two loops can be divided into two parts with different rates of change, as the current control loop has much faster dynamics than the comparatively slow voltage control loop. The current control loop has is called the fast boundary layer and the voltage loop is named the slow quasi-steady state. In order to analyze the stability of this cascade control loop, the dynamics of the system are separated into the slow dynamics and the fast dynamics.

According to (28) and **Figure 13**, the dynamics of the DC/DC converter for the current loop is given by:

$$\frac{di_L}{dt} = \frac{1}{L} v_{C_{in}} - \frac{1}{L} v_{DC}.d' \tag{29}$$

In (29), v_{DC}, $v_{C_{in}}$ are slow varying relative to the current and considered constants in this equation. In order to design the controller the following Lyapunov function is defined:

$$V_i = \frac{1}{2} e_i^2 \tag{30}$$

where $e_i = i_{ref} - i_L$. By considering (29) and (30), the following controller renders the derivative of the Lyapunov function negative definite:

$$d' = \frac{V_{Cin}}{V_{DC}} - k_p e_i - k_i \int e_i \tag{31}$$

According to **Figure 13**, the voltage loop includes the nonlinear droop block and the closed loop block of the inner current control loop. It is worthwhile to mention that the voltage loop does not try to render the voltage error zero. Thus, the asymptotic stability is not required and the boundedness of the voltage variables is the only requirement. In order to investigate the boundedness, the energy function for the voltage error signal is defined as:

$$V_v = \frac{1}{2} \times e_v^2 \tag{32}$$

where $e_v = V_{nom} - v_{DC}$.

According to (28) the dynamics of the voltage error is given by (due to the fast varying current loop, it is assumed that the inductor current and duty cycle have already reached their final values):

$$\frac{de_v}{dt} = \frac{1}{C_o.R_{eq}}(V_{nom} - e_v) - \frac{1}{C_o}.D'.I_L \tag{33}$$

The derivative of the energy function (32) is given by:

$$\dot{V}_v = \dot{e}_v \times e_v = \frac{1}{C_o.R_{eq}}\left[(V_{nom} - D'.I_L.R_{eq})e_v - e_v^2\right] \tag{34}$$

due to the fact that $(V_{nom} - D'.I_L.R_{eq})$ is close to zero the derivative of the energy function is negative. Thus the voltage error is bounded.

5. Simulation and experimental results

Simulation and experimental results are presented in this section. In **Tables 1** and **2** the specifications of the DC/DC converter used in the experimental prototype and simulations software are given in.

The DC MG configuration is shown in **Figure 3** which implemented to obtain the simulation results. The concept of dropbased control is based on reducing the voltage deviation caused by the change in MG load. Therefore, to simulate voltage deviation, two types of step change are applied in MG Load. The performance of the proposed control system is compared with the conventional drop-down method in reducing the MG voltage deviation. All the simulation results are implemented with $SoC = 80\%$. **Figure 14** shows the first type of step changes in DC MG loads in both charge/discharge modes. To illustrate and compare the performance of the proposed method with the conventional in term of power absorption and injection. As can be seen, the power absorption and induction of the proposed method is very significant in regulating the voltage which is presented in **Figure 15**. **Figure 16** shows a superior performance of the proposed method in voltage deviation by changing the DC MG load in both charge and discharge modes. **Figure 17** shows the second type of variations of the DC load inside MG for both charge/discharge modes. **Figures 18** and **19** shows the *SOC* value changes based on the type 2 load changes over the simulation time for the conventional and the proposed method, respectively. These *SOC* values have been used as the input of the optimization algorithm to obtain the simulation results. Each curve belongs to a specific value of the *SOC* (the *SOC* values have been varied from 20% to 80% to ensure the optimal operation of the storage systems).

Figures 20 and **21** show the DC-bus voltage variations for different values of *SOC* for the conventional and the proposed adaptive droop controllers, respectively.

Symbols	Parameters	Values
P_o	Nominal Output Power for Battery#1 converter	500 W
P_o	Nominal Output Power for Battery#2 converter	250 W
P_o	Nominal Output Power for PV converter	500 W
f_{sw}	Switching Frequency	100 kHz
V_{nom}	Nominal DC MG Voltage	190 V DC
$V_{in_{Bat}}$	Battery Bank Voltage	80 V DC
$V_{in_{PV}}$	PV Bank Voltage	80 V DC
L_1	Battery input Inductor	150μH
L_2	PV input Inductor	330μH
C_o	Output Capacitor	200μF

Table 2.
DC/DC converter specifications in experimental.

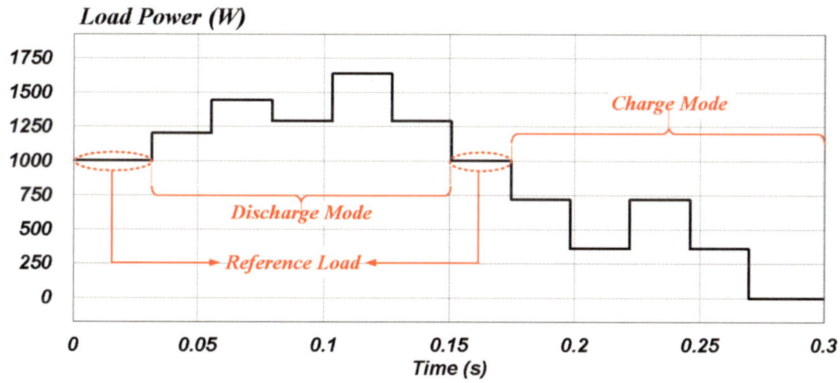

Figure 14.
Step changes (type 1) of the DC-load in DC MG.

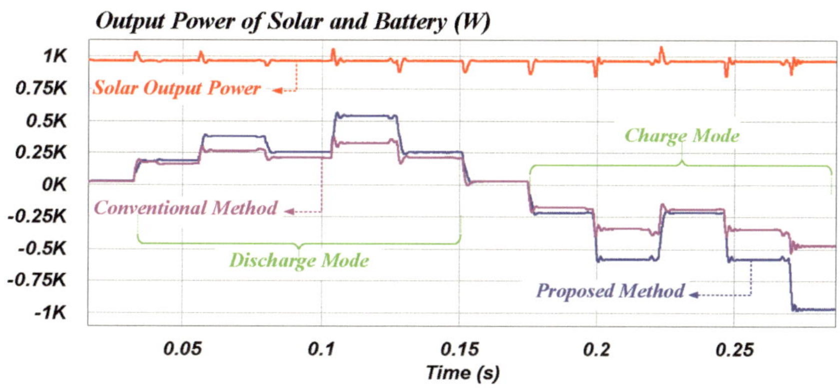

Figure 15.
Simulation result to compare the output power of the solar and battery between the proposed and conventional control method in SoC = 80%.

Figure 21 validates the superior performance of the proposed adaptive droop controller over the conventional one in regulating the DC-bus voltage for a wide range of *SOC* changes. In additions, the solar converter control system switches to the

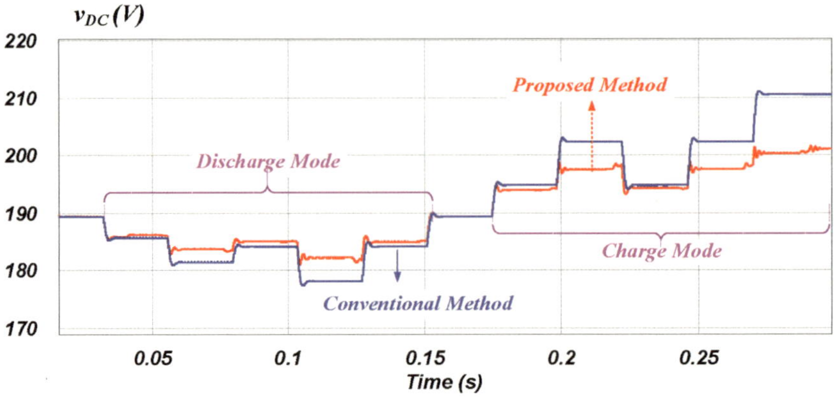

Figure 16.
Simulation result to compare the DC-bus voltage regulation in the battery between the proposed and conventional control method in SoC = 80%.

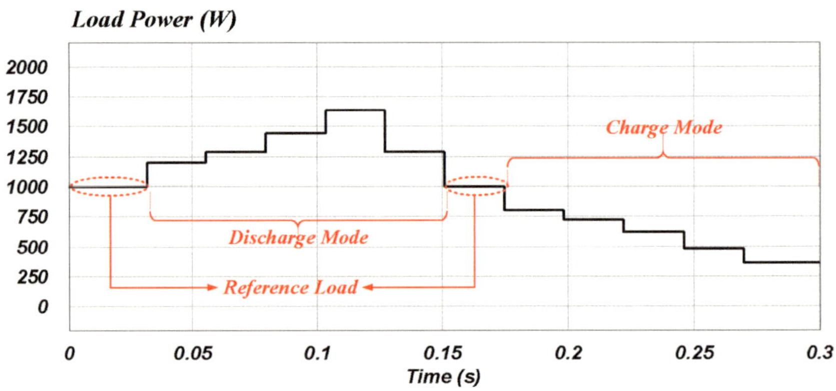

Figure 17.
Step changes (type 2) of the DC-load in DC MG.

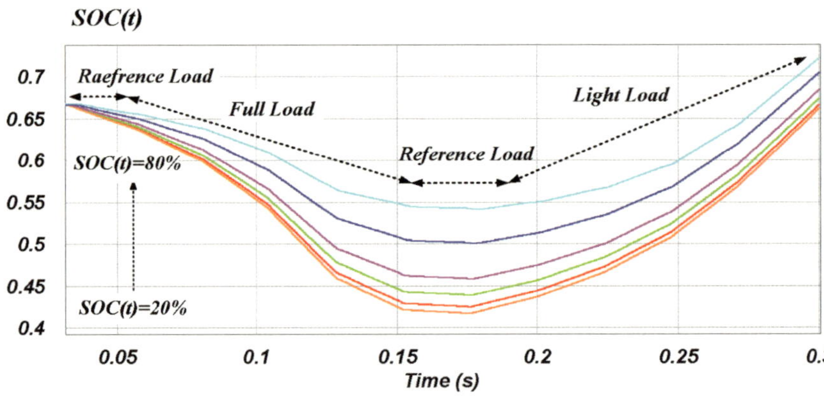

Figure 18.
Simulation results of SOC values when there are load step changes in the conventional control method.

droop control mode to fixed the DC voltage under 200 V in light load condition by state-modifier.

A small-scale DC micro-grid set-up has been implemented to evaluate the performance of the proposed method. The implemented experimental setup includes three

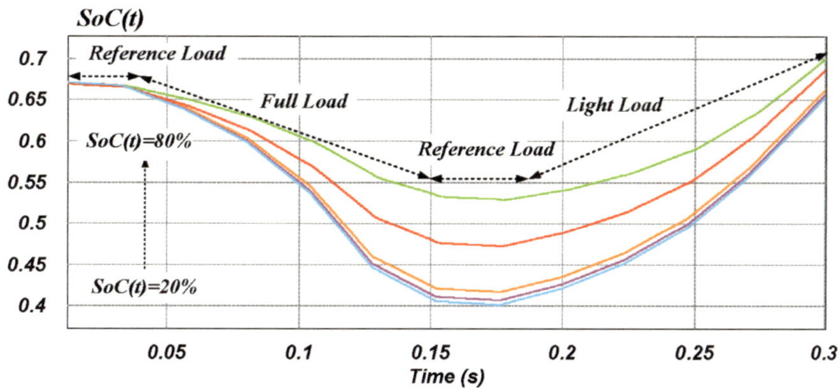

Figure 19.
Simulation results of SOC values for load step changes in the proposed control method.

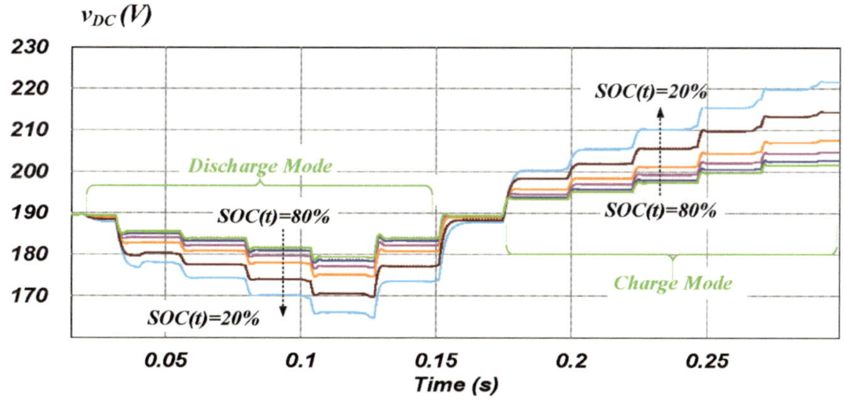

Figure 20.
Simulation result of the DC-bus voltage regulation of the battery by the load step changes in the conventional control method.

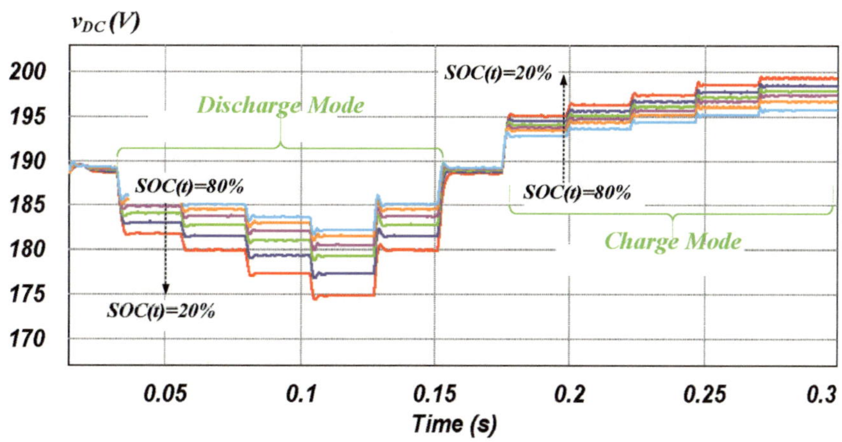

Figure 21.
DC-bus voltage regulation of the battery by the load step changes in the implemented SQP method in the proposed control method.

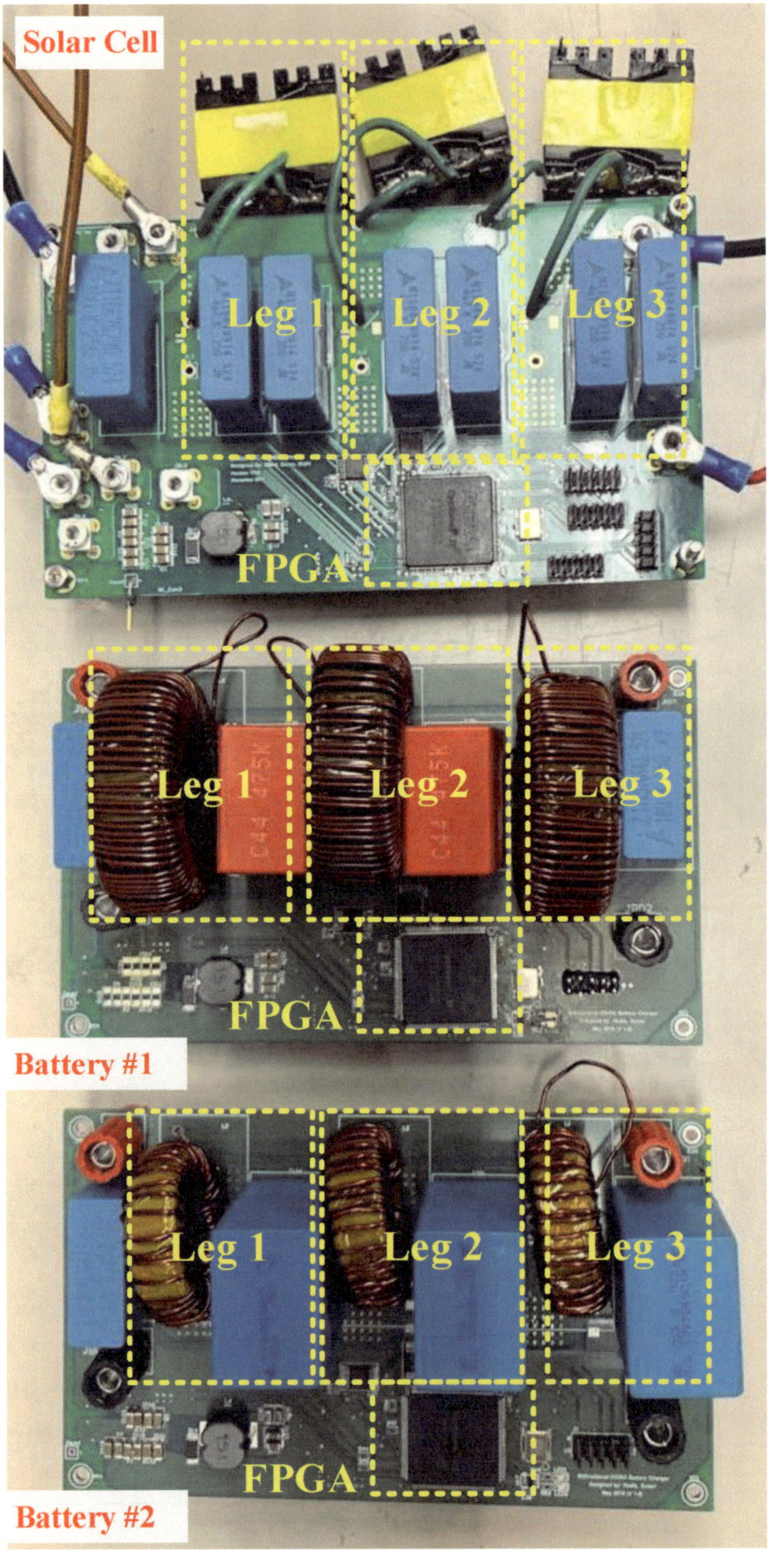

Figure 22.
Top side of the prototype DC/DC bidirectional-boost converters, the converters of the solar, battery one, and battery two.

boost converters in the power conditioning systems for the two energy storage (battery1 = 500 W and battery#1 = 250 W) and solar systems (PV panel = 500 W) with variation of DC loads. The specifications of the implemented converters are given in **Table 2**. **Figure 22** shows the experimental prototype of the bidirectional DC/DC converters, solar, battery#1, and battery#2 converters. The implemented converters use three separate phases for optimal performance (each phase changes phase 120 degrees to reduce the current wave). **Figure 23** shows a test prototype including two battery converters, a PV converter, a PV simulator, a battery simulator and a DC load to evaluate the proposed control system in the case of practical situation. For cost and safety reasons, experiments are performed at reduced power scale (1/2). All test results are based on a nominal voltage of 200V with a voltage variation of 40V. SOC values change from 20–80%, when load specifications change from reference load (500W) to the light load (50W) and from reference full load slow (1100W).

Figure 24 shows the high frequency wave forms of the implemented boost converter. **Figure 25** shows the performance of the PI controller in tracking the

Figure 23.
The test bench of two batteries and PV simulator with the DC load developed in the laboratory.

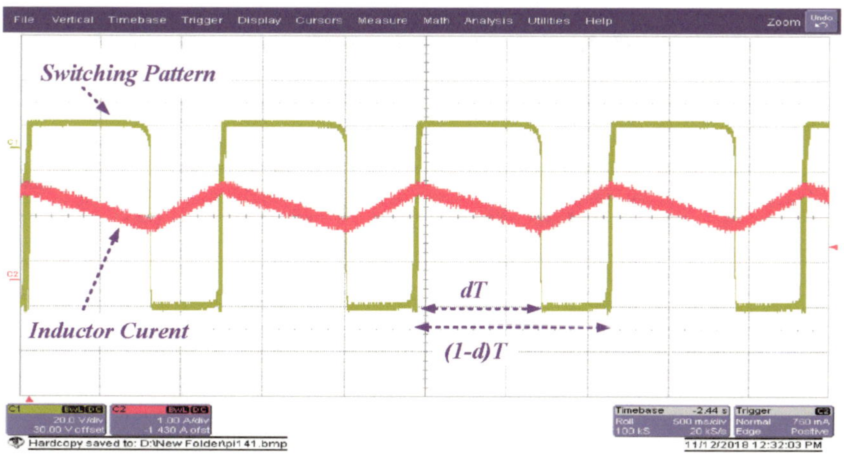

Figure 24.
Experimental result of the boost converter high frequency waveforms.

reference value of the output current. The proposed method is validated under light and heavy load changing conditions with $SoC = 40\%, 60\%$, and 80%. **Table 3** finalized the summary of the results obtained from the experimental setup. In **Table 3** the comparison between the proposed control system performance with the conventional droop controller performance is shown in terms of improving deviation in the DC-bus voltage as well as compensating the demanded load power. To evaluate the proposed system performance for a wide range of load variations in MG, two various cases with different step load changes considered in **Table 3**. Offering a superior performance over a wide range MG load a long with a wide range of available storage capacity (i.e, *SOC* of batteries) in comparison with the conventional droop is presented in **Table 3** for the proposed method. The above mentioned load changing strategies are listed as follows.

5.1 From reference to light load of $SoC = 80\%$

Figure 26 illustrates the performance of the conventional droop control when the load is changing from reference value to the light load (from 500 W to 50 W). The minimum voltage devotion in the conventional method is $17V$.

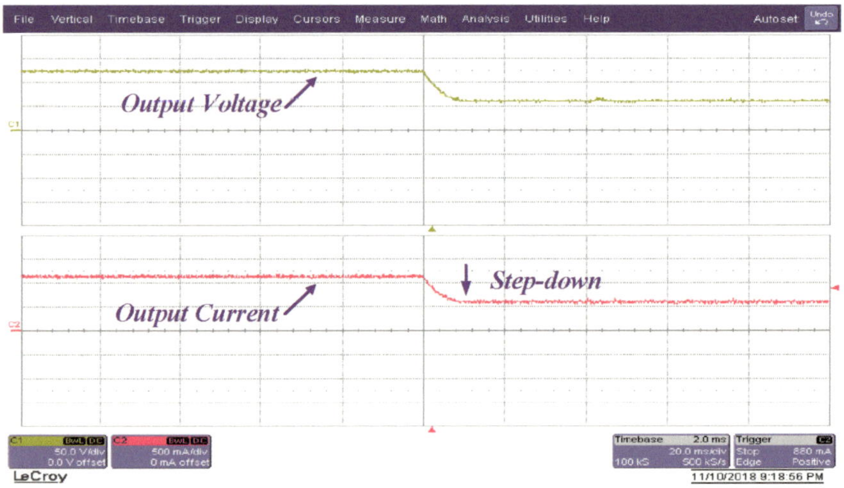

Figure 25.
Experimental result of the PI controller performance by creating a step change in the load value.

Load step change	Method	Batteries SOC value	Voltage deviation	Voltage deviation improvement*	Compensated load current	
					Battery #1	Battery #2
Reference to Light Load (500 W to 50 W)	Conventional	NA (80%)	17 V	NA (0%)	0.9A	0.42A
	Proposed	80%	7 V	58%	1.2A	0.65A
Reference to Full Load (500 W to 1100 W)	Conventional	NA (80%)	20 V	NA (0%)	1.1A	0.55A
	Proposed	80%	6 V	71%	1.8A	0.93A
		60%	11 V	45%	1.5A	0.75A
		40%	14 V	31%	1.3A	0.65A

Table 3.
Performance of the proposed control system in improving the DC-bus voltage deviation as well as compensating the load demanded power in comparison with the conventional droop controller.

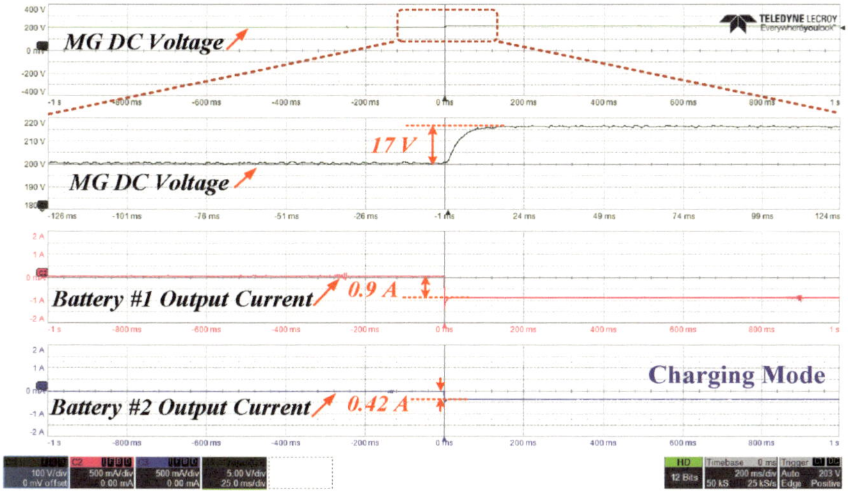

Figure 26.
Experimental result of the DC-bus voltage and batteries output current in conventional droop control method operating by one step change of load value from 500W to 50W, with SOC = 80%.

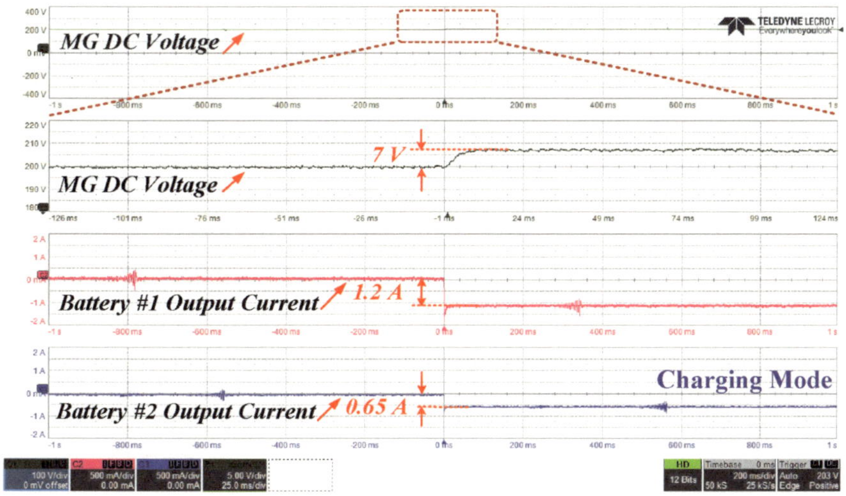

Figure 27.
Experimental result of the DC-bus voltage and batteries output current in proposed droop control method operating by one step change of load value from 500W to 50W, with SOC = 80%.

Figure 27 verifies the ability of the proposed controller in maintaining the DC-bus voltage against load steps for $SOC = 80\%$ in a lower voltage deviation, which voltage deviation is only $7V$. According to **Figure 27** and **Table 3**, the proposed controller could successfully compensate a higher percentages of load demanded power in comparison with the conventional method while offering a precise current sharing between two battery units.

5.2 From reference to full load of $SoC = 80\%$

Figure 28 presented the conventional droop control performance when the load changes from reference value to the full load (from $500W$ to $1100W$). The minimum voltage devotion in the conventional method is $6V$. **Figure 29** verifies the

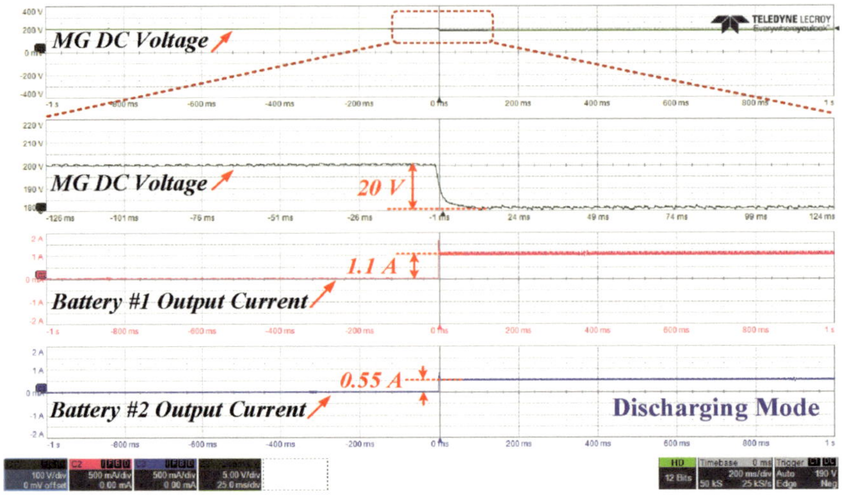

Figure 28.
Experimental result of the DC-bus voltage and batteries output current in conventional droop control method operating by one step change of load value from 500W to 1100W, with SOC = 80%.

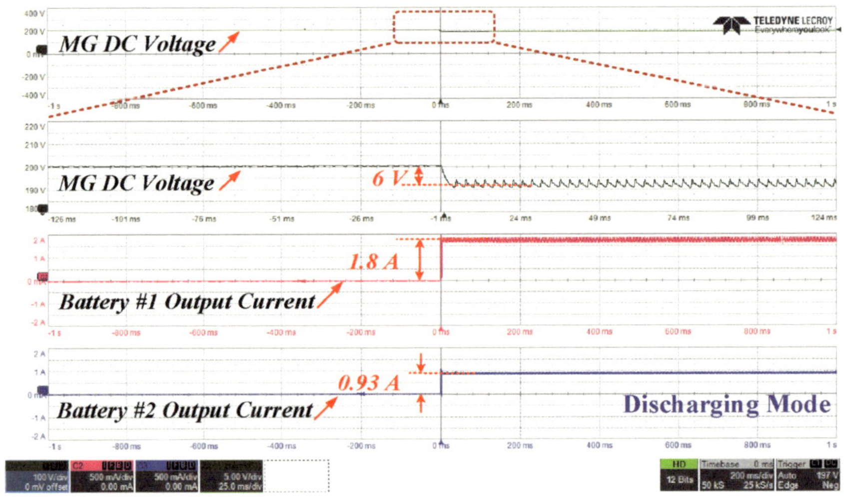

Figure 29.
Experimental result of the DC-bus voltage and batteries output current in proposed droop control method operating by one step change of load value from 500W to 1100W, with SOC = 80%.

ability of the proposed controller in maintaining the DC-bus voltage against load steps for $SOC = 80\%$ in a lower voltage deviation, which voltage deviation is only $6V$. According to **Figure 29** and **Table 3**, the proposed controller could successfully compensate a higher percentages of load demanded power in comparison with the conventional method while offering a precise current sharing between two battery units.

5.3 From reference to full load of $SoC = 80\%, 60\%,$ *and* 40%

Figures 29–31 illustrated the proposed control system performance in maintaining the DC voltage at three values of $SoC = 80\%, 60\%,$ and 40% in discharge mode. The voltage deviation in the three cases are respectfully, $6V$, $11V$, and

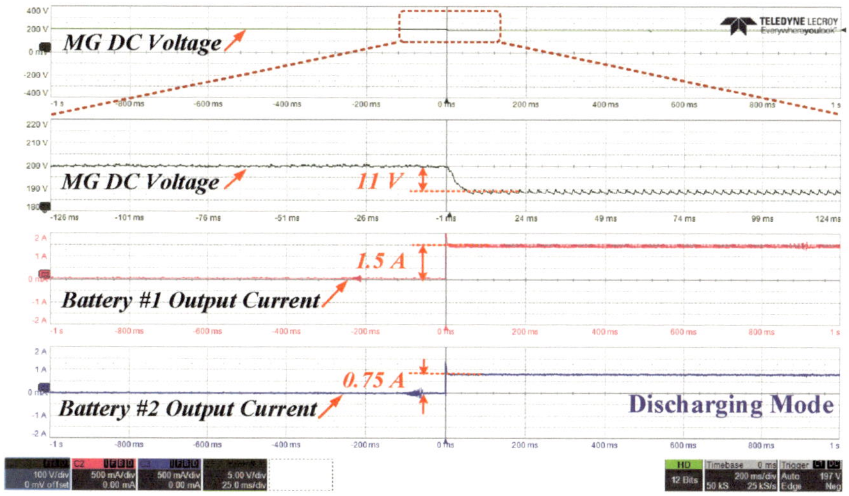

Figure 30.
Experimental result of the DC-bus voltage and batteries output current in proposed droop control method operating by one step change of load value from 500W to 1100W, with SOC = 60%.

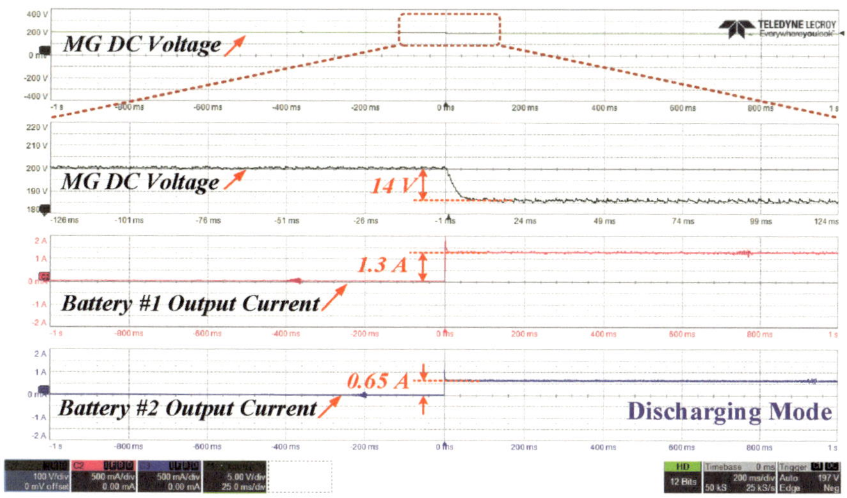

Figure 31.
Experimental result of the DC-bus voltage and batteries output current in proposed droop control method operating by one step change of load value from 500W to 1100W, with SOC = 40%.

$14V$. The results show that even in the worst case ($SoC = 40\%$) the performance of the proposed system is better than the conventional one.

Figure 32 shows the efficiency curve of the battery converter in the full range of output power variation from zero to full load. The rated power of the battery converter is 500 W as mentioned in **Table 2**. It should be mentioned that the proposed control system is a supervisory control system that control the flow of power between the converter and the DC grid by providing power references of the converter controller. Therefore, the proposed control system does not have any effect on the efficiency level of the converters and there are no extra losses caused by the proposed method. Moreover, the proposed method affects the control system only in transients and in steady-state operation, the control system is not affected by the proposed method.

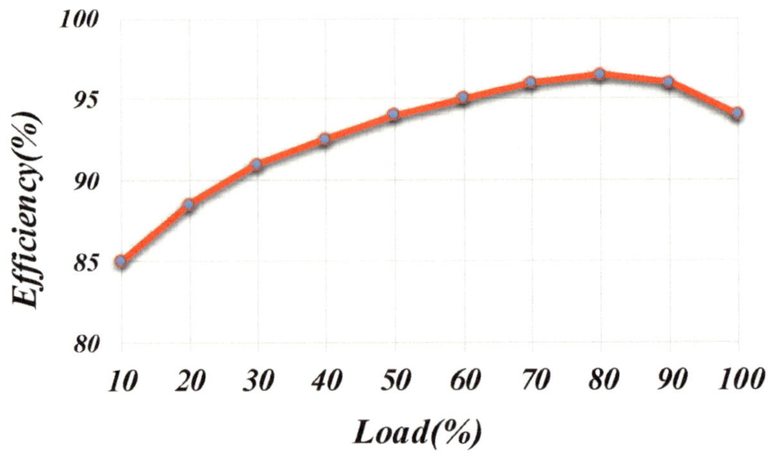

Figure 32.
Efficiency curves of the DC converter when the load is changing from light to full load.

6. Conclusion

In this chapter, a new adaptive droop control technique for the Battery or ESS has been proposed in the islanded mode. The proposed method provides a very tight DC-bus voltage regulation, while charge/discharge control task of the battery system is performed. In this control scheme, two extra operation modes (i.e., fast charge and fast discharge) has been proposed, which can offer an optimal performance for the DC MG. Thus, the extra power generated by the renewable energy sources can effectively charge the battery so fast. In case of voltage dropping, the battery is fast discharged, which can keep the desired DC-bus voltage range. In addition, a nonlinear optimization method has been introduced to determine the adaptive parameters in the optimal droop controller. Experimental and simulation results have validated the superior performance of the presented controller in comparison with the conventional controller.

Author details

Hadis Hajebrahimi*, Sajjad Makhdoomi Kaviri, Suzan Eren and Alireza Bakhshai
Department of Electrical and Computer Engineering, Queen's University, Kingston, ON, Canada

*Address all correspondence to: hadis.hajebrahimi2@gmail.com

References

[1] F. Blaabjerg, Zhe Chen and S. B. Kjaer, "Power electronics as efficient interface in dispersed power generation systems," in IEEE Transactions on Power Electronics, vol. 19, no. 5, pp. 1184-1194, Sept. 2004.

[2] D. E. Olivares, A. Mehrizi-Sani, A. H. Etemadi, C. A. Canizares, R. Iravani, M. Kazerani, A. H. Hajimiragha, O. Gomis-Bellmunt, M. Saeedifard, R. Palma-Behnke, and N. D. Hatziargyriou, "Trends in Microgrid control," IEEE Transactions on Smart Grid, vol. 5, no. 4, pp. 1905–1919, Jul.

[3] J. M. Guerrero, M. Chandorkar, T. L. Lee and P. C. Loh, "Advanced Control Architectures for Intelligent Microgrids—Part I: Decentralized and Hierarchical Control," in IEEE Transactions on Industrial Electronics, vol. 60, no. 4, pp. 1254-1262, April 2013.

[4] J. M. Guerrero, J. C. Vasquez, J. Matas, L. G. de Vicuna and M. Castilla, "Hierarchical Control of Droop-Controlled AC and DC Microgrids—A General Approach Toward Standardization," in IEEE Transactions on Industrial Electronics, vol. 58, no. 1, pp. 158-172, Jan. 2011.

[5] H. Han, X. Hou, J. Yang, J. Wu, M. Su and J. M. Guerrero, "Review of Power Sharing Control Strategies for Islanding Operation of AC Microgrids," in IEEE Transactions on Smart Grid, vol. 7, no. 1, pp. 200-215, Jan. 2016.

[6] Y. Han, H. Li, P. Shen, E. A. A. Coelho and J. M. Guerrero, "Review of Active and Reactive Power Sharing Strategies in Hierarchical Controlled Microgrids," in IEEE Transactions on Power Electronics, vol. 32, no. 3, pp. 2427-2451, March 2017.

[7] F. Wang, Y. Pei, D. Boroyevich, R. Burgos, and K. Ngo, "Ac vs. Dc Distribution for Off-Shore Power Delivery," in 34th Annual Conference of IEEE Industrial Electronics (IECON), 2018, pp. 2113–2118.

[8] J. Park and J. Candelaria, "Fault Detection and Isolation in Low-Voltage DC-Bus Microgrid System," in IEEE Transactions on Power Delivery, vol. 28, no. 2, pp. 779-787, April 2013.

[9] L. Meng, Q. Shafiee, G. Ferrari Trecate, H. Karimi, D. Fulwani, X. Lu, and J. M. Guerrero, "Review on Control of DC Microgrids, â" IEEE J. Emerg. Sel. Top. Power Electron., pp. 1–1, 2017.

[10] E. Rodriguez-diaz, J. C. Vasquez, and J. M. Guerrero, "Intelligent DC homes in future sustainable energy systems: When efficiency and intelligence work together," IEEE Consum. Electron. Mag., vol. 5, no. 1, pp. 74–80, 2016.

[11] A. Khorsandi, M. Ashourloo, and H. Mokhtari, "A decentralized control method for a low-voltage DC microgrid," IEEE Trans. Energy Convers.vol. 29, no. 4, pp. 793–801, Dec. 2014.

[12] Y. Han, H. Li, P. Shen, E. A. A. Coelho and J. M. Guerrero, "Review of Active and Reactive Power Sharing Strategies in Hierarchical Controlled Microgrids," in IEEE Transactions on Power Electronics, vol. 32, no. 3, pp. 2427-2451, March 2017.

[13] Y. Gu, X. Xiang, W. Li and X. He, "Mode-Adaptive Decentralized Control for Renewable DC Microgrid with Enhanced Reliability and Flexibility," in IEEE Transactions on Power Electronics, vol. 29, no. 9, pp. 5072-5080, Sept. 2014.

[14] S. Augustine, M. K. Mishra, and N. Lakshminarasamma, "Adaptive droop control strategy for load sharing and circulating current minimization in low-

voltage standalone DC microgrid," IEEE Trans. Sustain. Energy, vol. 6, no. 1, pp. 132–141, Jan. 2015.

[15] T. Dragicevic et al., "DC Microgrids; Part I: A Review of Control Strategies and Stabilization Techniques,"IEEE Trans. Power Electron., vol. 31, no. 7, July 2016, pp. 4876–4891.

[16] F. Chen, R. Burgos, D. Boroyevich, J. C. Vasquez and J. M. Guerrero, "Investigation of Nonlinear Droop Control in DC Power Distribution Systems: Load Sharing, Voltage Regulation, Efficiency, and Stability," in IEEE Transactions on Power Electronics, vol. 34, no. 10, pp. 9404-9421, Oct. 2019.

[17] P. Prabhakaran, Y. Goyal and V. Agarwal, "Novel Nonlinear Droop Control Techniques to Overcome the Load Sharing and Voltage Regulation Issues in DC Microgrid," in IEEE Transactions on Power Electronics, vol. 33, no. 5, pp. 4477-4487, May 2018.

[18] Chen, D and Xu, L 2017, 'AC and DC microgrid with distributed energy resources' in Technologies and Applications for Smart Charging of Electric and Plug-in Hybrid Vehicles. Springer, pp. 39-64., 10.1007/978-3-319-43651-7-2.

[19] X. Lu, K. Sun, J. M. Guerrero, J. C. Vasquez, and L. Huang, "State-of-charge balance using adaptive droop control for distributed energy storage systems in DC microgrid applications," IEEE Trans. Ind. Electron., vol. 61, no. 6, pp. 2804–2815, Jun. 2014.

[20] Dragicevic, T., Lu, X., Vasquez, J., et al.: 'DC microgrids – part II: a review of power architectures, applications, and standardization issues', IEEE Trans. Power Electron., 2016, 31, (5), pp. 3528–3549.

[21] Mahmood, H., Michaelson, D., Jiang, J.: 'Decentralized power management of a PV/battery hybrid unit in a droop controlled islanded microgrid', IEEE Trans. Power Electron., 2015, 30, (12), pp. 7215–7229

[22] J. M. Guerrero, J. C. Vasquez, J. Matas, L. G. de Vicuna, and M. Castilla, "Hierarchical control of droop-controlled AC and DC microgrids—A general approach toward standardization," IEEE Trans. Ind. Electron., vol. 58, no. 1, pp. 158–172, Jan. 2011.

[23] IEEE 929–2000 Standard, "IEEE Recommended Practice for Utility Interface of Photovoltaic (PV) Systems", April 2000.

[24] A. Kwasinski, "Quantitative evaluation of DC microgrids availability: Effects of system architecture and converter topology design choices," IEEE Trans. Power Electron., vol. 26, no. 3, pp. 835–851, Mar. 2011.

[25] D. Jovcic, M. Taherbaneh, J. Taisne, and S. Nguefeu, "Offshore DC grids as an interconnection of radial systems: protection and control aspects," IEEE Trans. Smart Grid, vol. 6, no. 2, pp. 903–910, 2015.

[26] J. Park and J. Candelaria, "Fault Detection and Isolation in Low-Voltage DC-Bus Microgrid System,"in IEEE Transactions on Power Delivery, vol. 28, no. 2, pp. 779-787, April 2013.

[27] Q. Shafiee, J. M. Guerrero and J. C. Vasquez, "Distributed Secondary Control for Islanded Microgrids - A Novel Approach," in IEEE Transactions on Power Electronics, vol. 29, no. 2, pp. 1018-1031, Feb. 2014.

[28] E. Barklund, N. Pogaku, M. Prodanovic, C. Hernandez-Aramburo and T. C. Green, "Energy Management in Autonomous Microgrid Using Stability-Constrained Droop Control of Inverters," in IEEE Transactions on Power Electronics, vol. 23, no. 5, pp. 2346-2352, Sept. 2008

[29] Y. Karimi, H. Oraee, M. S. Golsorkhi and J. M. Guerrero, "Decentralized Method for Load Sharing and Power Management in a PV/Battery Hybrid Source Islanded Microgrid," in IEEE Transactions on Power Electronics, vol. 32, no. 5, pp. 3525-3535, May 2017.

[30] E. Barklund, N. Pogaku, M. Prodanovic, C. Hernandez-Aramburo and T. C. Green, "Energy Management in Autonomous Microgrid Using Stability-Constrained Droop Control of Inverters," in IEEE Transactions on Power Electronics, vol. 23, no. 5, pp. 2346-2352, Sept. 2008

[31] N. L. Diaz, T. Dragičević, J. C. Vasquez, and J. M. Guerrero, "Intelligent distributed generation and storage units for DC microgrids—A new concept on cooperative control without communications beyond droop control," IEEE Trans. Smart Grid, vol. 5, no. 5, pp. 2476–2485, Sep. 2014.

[32] Y. Gu, X. Xiang, W. Li, and X. He, "Mode-adaptive decentralized control for renewable DC microgrid with enhanced reliability and flexibility," IEEE Trans. Power Electron., vol. 29, no. 9, pp. 5072–5080, Sep. 2014.

[33] E. M. L. Beale, Numerical Methods6in: Nonlinear Programming, J. Abadie, ed., NorthHolland, Amsterdam, 1967.

[34] Powell, M. J. D. "A Fast Algorithm for Non-linearly Constrained Optimization Calculations." Numerical Analysis, ed. G. A. Watson, Lecture Notes in Mathematics, Springer-Verlag, Vol. 630, 1978

[35] K. Wu, C. W. de Silva and W. G. Dunford, "Stability Analysis of Isolated Bidirectional Dual Active Full-Bridge DC–DC Converter With Triple Phase-Shift Control," in IEEE Transactions on Power Electronics, vol. 27, no. 4, pp. 2007-2017, April 2012.

Section 3

Power Quality in Micro-Grids with Energy Storage Technologies

Chapter 3

Power Quality in Renewable Energy Microgrids Applications with Energy Storage Technologies: Issues, Challenges and Mitigations

Emmanuel Hernández Mayoral, Efraín Dueñas Reyes, Reynaldo Iracheta Cortez, Carlos J. Martínez Hernández, Carlos D. Aguilar Gómez, Christian R. Jiménez Román, Juan D. Rodríguez Romero, Omar Rodríguez Rivera, Edwin F. Mendoza Santos, Wilder Durante Gómez and José I. Barreto Muñoz

Abstract

Nowadays, the electric power distribution system is undergoing a transformation. The new face of the electrical grid of the future is composed of digital technologies, renewable sources and intelligent grids of distributed generation. As we move towards the electrical grid of the future, microgrids and distributed generation systems become more important, since they are able to unify small-scale and flexible generation to clean energy and intelligent controls. The microgrids play an important role in marking electrical grids more robust in the face of disturbances, increasing their resilience. Although the microgrid concept continues in discussion in technical circles, it can be defined as an aggregation of electrical elements in low generation voltage, storage and loads (users) which are grouped in a certain bounded geographical area. The issues of a microgrid integrated with energy storage technologies has gained increasing interest and popularity worldwide as these technologies provide the reliability and availability that are required for proper operation in the system. Actual studies show that the implementation of energy storage technologies in a microgrid improves transients, capacity, increases instantaneous power and allows the introduction of renewable energy systems. However, there are still certain unsolved problems in power quality terms. This article clearly describes those problems generated by each storage technology foe microgrids applications. All the ideas in this review contribute significantly to the growing effort towards developing a cost-effective and efficient energy storage technology model with a long-life cycle for sustainable implementation in microgrids.

Keywords: Distributed generation, microgrids, energy storage systems, power quality, renewable energy sources

1. Introduction

Energy storage systems (ESS) and their microgrids application play a very important role in the electricity industry since they mitigate the problem of intermittency of renewable energy sources (RES) [1–4] while improving stability of the microgrid performing auxiliary services such as the decrease in demand at peak hours, protection against blackouts and control of power quality [5–7]. ESS also help renewable energy integration by managing the energy balance during an energy crisis, therefore system stability has a significant effect on the overall electrical system by storing energy during off-peak hours at reduced cost [8–15]. Additionally, ESS can be applied for cases of energy arbitrage [16], decrease in demand at peak hours [17], load flow [18], spinning reserve [19], voltage support and regulation [20], black–start [20, 21], frequency regulation [7], power quality [22, 23], power reliability [24], changes in RES [25, 26], transmission and distribution systems modernization [27], electrical congestion mitigation [28], and off-grid services [25, 28]. That is why ESS have become widely used solutions [29–31]. In fact, to enhance the ESS capacities required by microgrid, a hybrid solution is commonly adopted [32]. However, there are still challenges in the ESS implementation for microgrid applications such as the adequate management of these technologies, power electronics, energy conversion mechanisms, reliability and some problems with the power quality derived from the intermittency of RES which affect the system frequency. To counteract these drawbacks, different solutions have been proposed that will be described in later sections where they not only improve it but also efficiently solve problems related to power control, voltage stability and the power factor.

Microgrid is defined, according to the US Department of Energy, as a group of loads, micro-sources and distributed energy resources with clearly defined electrical limits capable of being self-sufficient and operating autonomously from the distribution grid in order to ensure the continuity of the electricity supply with a high reliability factor [33]. Another microgrid concept, according to the Consortium for Electrical Reliability Technology Solutions (CERTS), is that of an entity consisting of distributed energy resources, as well as controllable electrical and thermal loads. These loads are connected to the upstream grid for power generation through photovoltaic panels, wind plants, fuel cells, diesel generators and micro-turbines with ESS [34] as seen in **Figure 1**. Simply put, an microgrid is a miniature version of the sustainable energy model that can be used to generate, distribute and control bi-directional energy flow within its operating limits in a coordinated, intelligent and efficient manner, with a focus on renewable energies integration. Microgrids can be connected and disconnected from the main grid to allow it to operate in both "grid connected mode" [35] and "island mode" [36]. Microgrid must have flexible characteristics in its operation in both modes of operation to improve the efficiency and security of the grid [37]. When the microgrid operate in "connected grid mode" can maintain a stable system frequency by exchanging power with the main grid. However, in "island mode", the microgrid are designed as off-grid systems [38] where primary frequency control is critical. Nevertheless, "island mode" is the most prominent feature of a microgrid, which is enabled through the use of switches at the point common coupling (PCC), which allows the microgrid to disconnect from the grid in case of upstream disturbances or voltage fluctuations [39]. Microgrid comprises only a portion of the distribution grid (generally in low–voltage), located next to the substation which contains a set of electrical and/or thermal loads, DG different types, distributed storage technologies with distinct features and capabilities.

Basically, microgrids offer significant benefits for both users and the electrical grid, reducing carbon emissions through the RES diversification, economic operation by reducing transmission and distribution costs (T&D), use of DG sources less expensive, energy efficiency responding to market prices in real–time, and better power quality

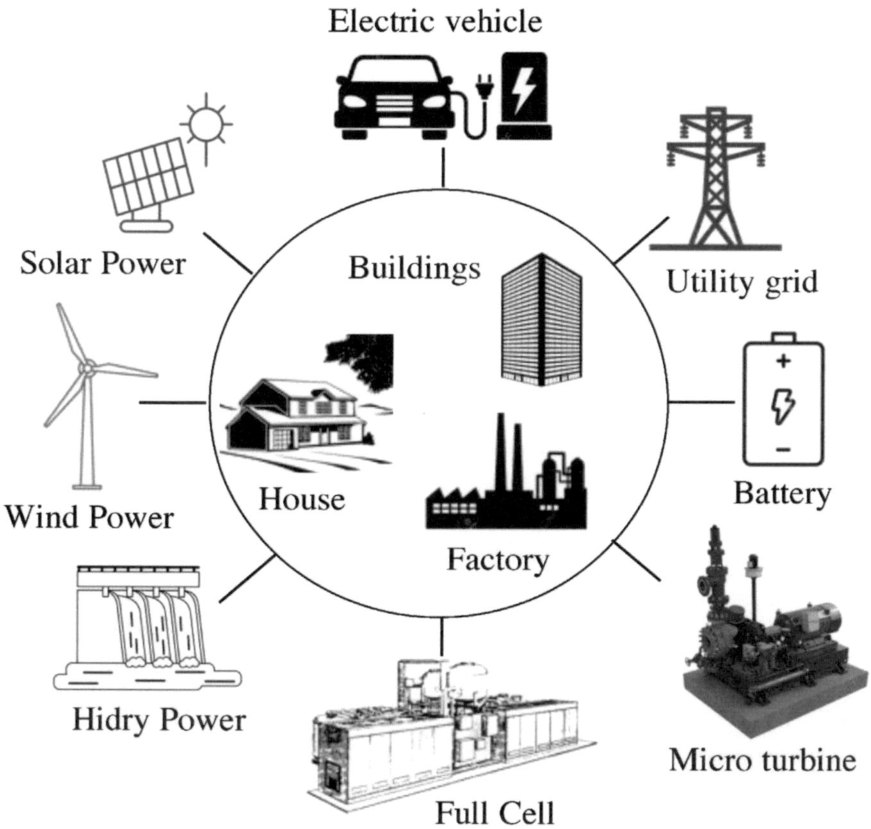

Figure 1.
MG typical structure.

when managing local loads. Therefore, the objective of this review is to pre-sent the actual state of ESS and their microgrid application in terms of power quality being the main contribution of this document: an ESS critical evaluation, highlighting their operational characteristics by minimizing the risk of supply interruptions, optimizing the consumption curve and reducing the maximum power required, which generates significant economic savings in the fixed term of generated power.

2. Microgrids overview

Microgrids can be classified as: AC–microgrid, DC–microgrid, and hybrid micro-grid (DC–AC microgrid).

2.1 AC–microgrid

A typical AC–microgrid is shown in **Figure 2**. In this system, all DG that include storage devices and loads are linked to the busbars of the AC mains by an electronic power converter. However, it is possible to connect AC generators, such as micro-turbines, diesel and wind turbines, directly to the main grid without the need for converters. Alternatively, to connect DC power sources such as batteries and PV systems to the grid, a DC/AC inverter is essential. Therefore, the loads are con-nected in a straight line to the AC bus bar. However, AC–microgrids have several drawbacks and such a grid involves complex control and synchronization problems. However, this grid is still widely used today [40]. An important detail to mention is

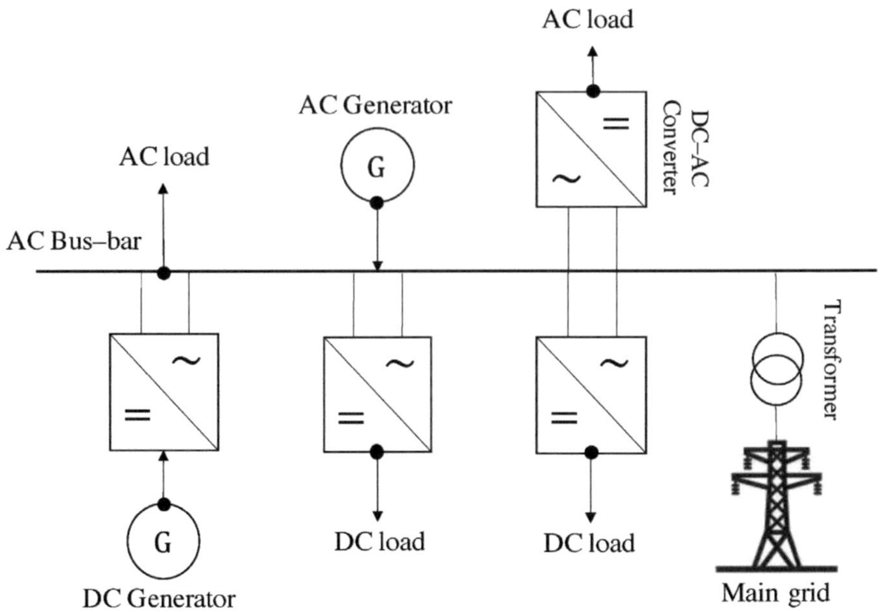

Figure 2.
AC–microgrid typical structure.

that the regulation of power quality in an AC–microgrid is carried out based on the conventional distribution system and the mode of operation [41].

2.2 DC–microgrid

Most of the generators that make up an microgrid produce DC power, which must be converted to AC power to accommodate the main grid. For which, it is required to perform the DC conversion at the end of the system since some equipment requires AC power to operate. However, converting DC/AC/DC power into an AC–microgrid reduces efficiency and causes power losses. This can be remedied by using high voltage DC operation as a benchmark, as the DC–microgrid is designed to address this problem. **Figure 3** shows the structure of a DC–microgrid. Unlike an AC–microgrid, the DC–microgrid offers considerable energy savings by reducing the number of converters in a single conversion process using a single converter. The authors in [41] stated that DC–microgrid are more suitable for distribution systems in residential areas than AC distributed networks causing few power quality problems. One of the best advantages of DC–microgrid is that they solve some control problems in the microgrid, making DG timing no longer necessary and ensuring that the controls are highly dependent on the DC bus voltage. Furthermore, the primary control is considerably simpler due to the absence of reactive power flow management. Also, many modern devices are DC powered and do not have power electronics that generate harmonics. Consequently, the level of conversion in DC–microgrid is low because it skips the CA stage in the middle of the process [42]. As a conclusion to this section, the operation of a DC–microgrid is smoother than AC–microgrid since phase and frequency monitoring are not taken into account [43].

2.3 Hybrid microgrid (DC–AC microgrid)

Hybrid–microgrid consist of AC and DC grids interconnected by large-scale multi-directional converters. This system could decrease the conversion stages

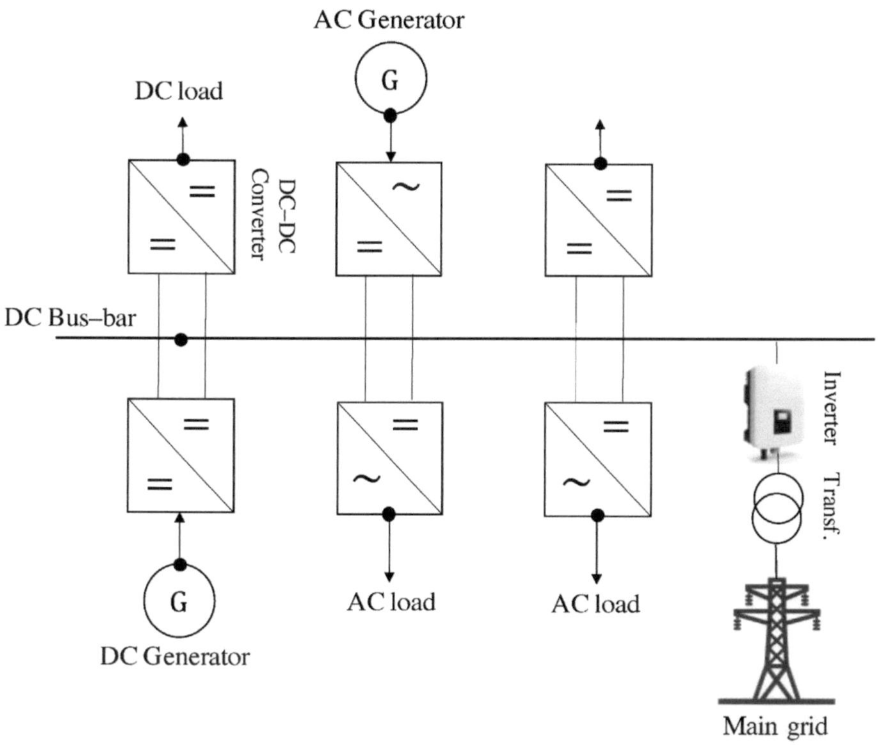

Figure 3.
DC–microgrid typical structure.

(DC/AC/DC and AC/DC/AC) into individual DC–microgrids or AC–microgrids and thus reduce the occurrence of power quality issues. In these types of microgrids, the AC sources and loads are tied to the AC bus, while the DC sources and loads are tied to the DC bus. The storage system can be linked to either of the two microgrids. **Figure 4** illustrates the one-line diagram of a hybrid microgrid [44, 45]. In a hybrid microgrid, the grid-connected mode of operation will supply or use the power from the main grid to meet power generation and load demand requirements. When disturbances arise, the microgrid must isolate itself from the main grid and work in autonomous mode. In grid-connected mode of operation, the microgrid operates efficiently to ensure critical load delivery is not compromised. The transient that occurs during the switching phase must be well controlled to avoid destroying the devices in the microgrid. Therefore, power quality issues need further investigation in this case [46].

3. ESS advances in microgrid applications

ESS are classified as: mechanical, electrochemical, electrical, thermal, and hybrid. Furthermore, these systems can be classified according to the formation process and the materials used, such as batteries [47], compressed air [48], flywheels [49], super-capacitors [50], superconducting magnetic energy storage (SMES) [51], fuel cells [52] and hybrid storage [53–55], which, the latter, are the most widely applied in micro-grids. These systems will be discussed in more detail below.

3.1 Batteries

Batteries store energy in an electrochemical form, and are available in different sizes and capacities ranging from 100 W to several MW. Batteries overall estimated

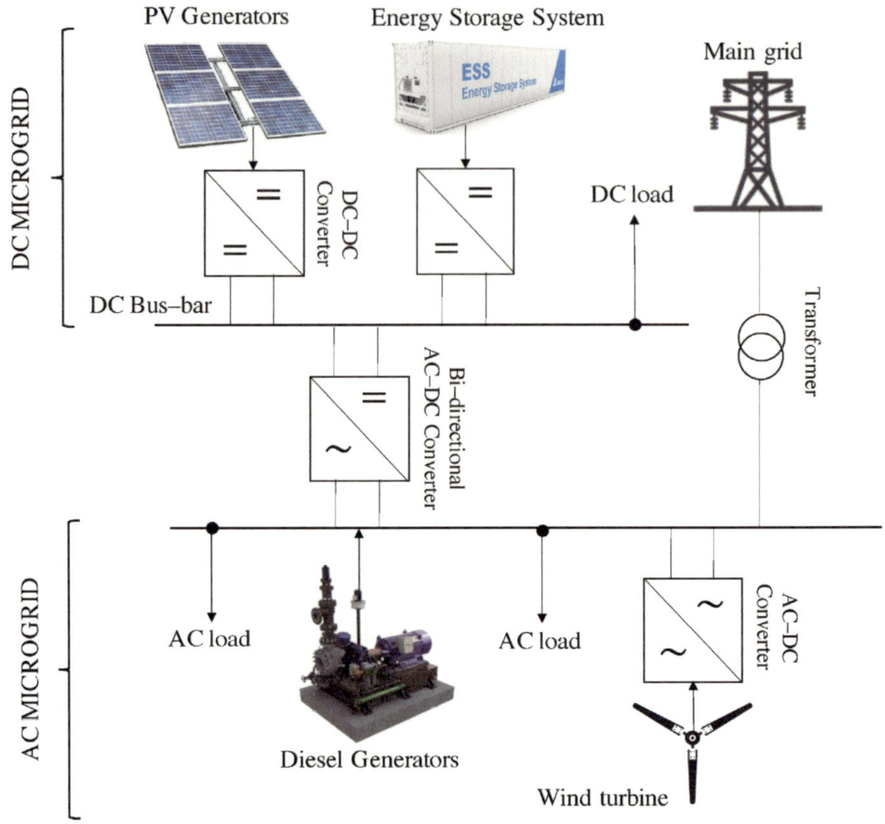

Figure 4.
Hybrid microgrid typical structure.

efficiency is in the range of 58–85%, depending on the operating cycle and the type of electrochemistry within the batteries. Lead-acid, Ni-Fe, Ni-Cd, Ni-M hydride, and Li-ion batteries are the five main types of energy storage based on batteries for microgrid applications. **Figure 5** shows, schematically, a constant increase in the energy density of batteries over the years. Lead-acid battery is the most technologically mature and lowest-cost energy storage device of all available battery technologies. However, the limited charge cycle capacity of these batteries typically results in an unacceptable scenario in system economics. On the other hand, Ni-Cd and Ni-M hydride batteries offer potential advantages over lead-acid batteries as they are environmentally friendly and provide a life cycle equivalent to that of lead-acid batteries, and an increase in its capacity (between 25 and 40%). As for the Li-ion battery, it has the highest energy density, but its cost is very high [56]. From a techno-economic aspect, Ni-M hydride battery is potentially the most competent technology in terms of: output power, voltage profile and charge–discharge characteristics, while the lead-acid battery turns out to be the most economical for renewable energy applications compared to Ni-Cd, Ni-M hydride, and Li-ion batteries. In general terms, due to their long service life and relatively low costs, but with a slow response, these types of batteries are ideal for applications with low duty cycles.

A microgrid composed of RES connected by electronic power converters can experience difficulties due to the voltage and current harmonics presence. These currents can, in turn, because voltage drops in line impedances. Additionally, voltage fluctuations and harmonic distortion can cause problems such as equipment tripping, overheating, and system malfunction. Microgrid stability depends on the

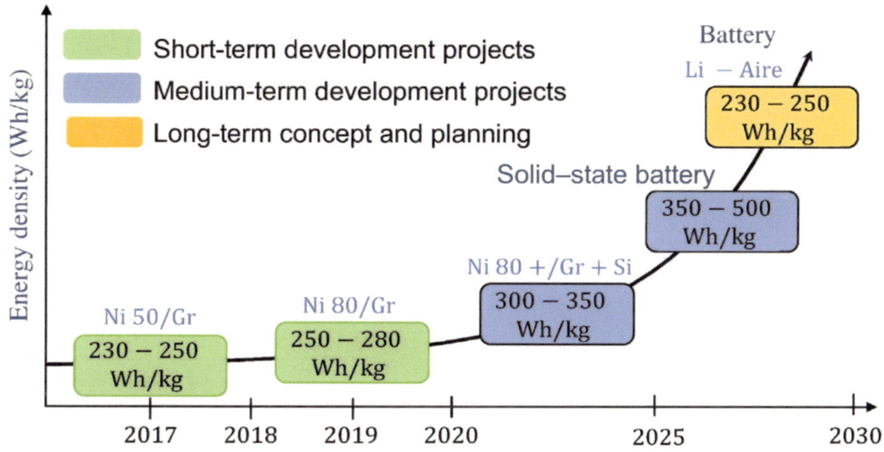

Figure 5.
Increased energy density of batteries.

ability of its units to mitigate and compensate for these phenomena. It should be considered that the batteries used to simultaneously exchange active power between the same battery and the main grid will significantly improve the power quality of the microgrid. This can be done by independent cascade control of currents and active and reactive power, thus controlling the reactive power balance and thus ensuring voltage stability across the microgrid. It should be noted that batteries are also used as an active harmonic filter. In addition to the above, batteries have the ability to maintain the voltage and frequency of the microgrid within the limits prescribed by the standards, since it can provide frequency support approximately 100 times faster than conventional generators. Finally, the batteries can withstand long-term voltage variations due to higher energy density and, therefore, will considerably improve the power quality in the microgrid.

3.2 Flywheels

A flywheel stores electrical energy in the form of kinetic energy and can convert the kinetic energy, back, to electrical energy when required. The energy stored in the flywheels is usually extracted from an electrical source from the grid or from any other source of electrical energy. When the flywheel is accelerated it stores energy and decelerates when discharge, to deliver the accumulated energy. The rotating flywheel is driven by an electric machine (electric motor-generator) that performs the exchange of electric energy to kinetic energy and vice versa. The flywheel and the electric machine have a common axis of rotation, so the control of the electric machine makes it possible to control the flywheel. This flywheel consists of a massive rotating cylinder (disk) that is supported on a stator by magnetic levitation bearings [57] as seen in **Figure 6**. It is divided into two categories: low speed, that is, from 6×10^3 to $\times10^5$ rpm (high inertia and low speed) with a mixed gearbox which provides an energy boost to short term (10–30 s) and which are the most popular in the industry [58]. The high speed, that is, $\times10^5$ (low inertia and high speed) that use a magnetic gearbox which are used in the aerospace industry [59]. Therefore, as the rotation speed of the flywheel rotor increases, the stored energy also increases proportionally with respect to the angular momentum. This stored energy can be used to rotor torque decelerate (discharge mode) by returning the kinetic energy to the electric motor, which acts as a generator. The nominal power can reach 52 MW, with storage capacities in the range of 3–148

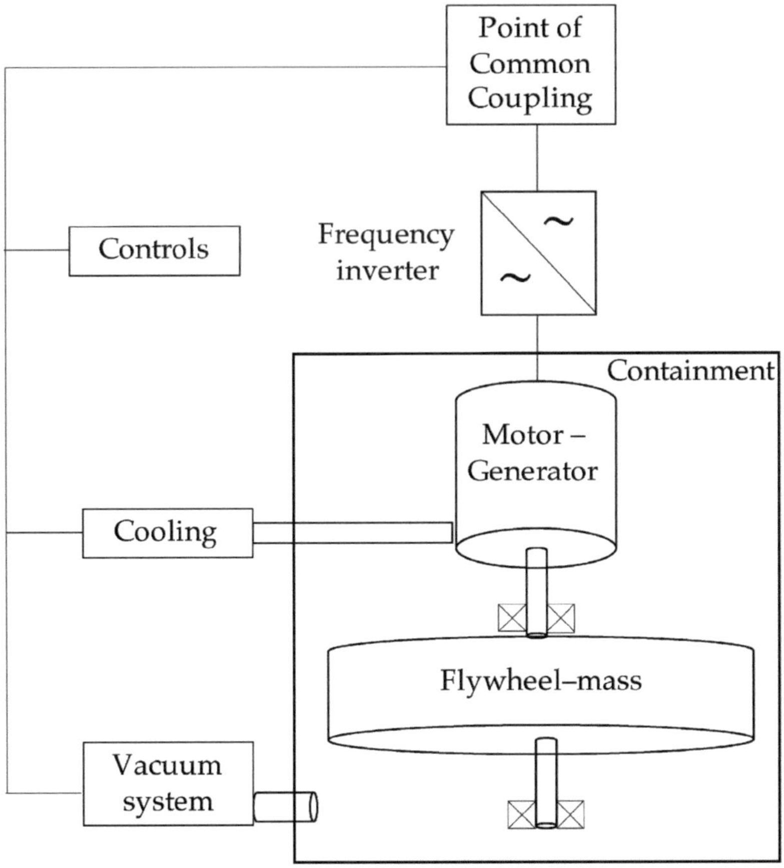

Figure 6.
Flywheel basic structure.

kWh. These flywheels present a self-discharge of between 2.8–21.9% per hour, with efficiencies of 88 to 96%. They have 20,000 charge and discharge cycles. Response time is milliseconds and discharge time is seconds to no more than 1 hour. Rapid charging of a system occurs in less than 15 minutes [60]. Compared to batteries, flywheels can perform better when a sudden energy deficiency occurs in the electricity generation from RES (solar or wind) [61]. A very important aspect to consider in the implementation of this type of technology is its low maintenance cost ($22 dollars/kW-year) although the acquisition cost is generally high ($5000 dollars/kWh). Considering the above, the flywheels become reliable and friendly devices with the CO_2 emissions reduction.

To connect the microgrid to the main grid and make it available to loads, the power quality must meet the established requirements. As part of those requirements, the frequency and voltage of the system must be kept at an acceptable level without deviations. However, voltage drop has become one of the main power quality problems that affects sensitive loads, increases line losses, increases neutral conductor overloads, and increases rotation losses in AC drives frequency. About 92% of power quality problems in microgrids are due to voltage dips and 80% of these last only 20–50 ms. Flywheels, given their excellent characteristics, can be a viable alternative to counteract this phenomenon, which can quickly add or extract power from the grid, to keep the system voltage and frequency within the acceptable range. Flywheels can also aid the penetration of wind and solar energy into electrical power systems, improving their stability. The fast response characteristics of flywheels make them suitable in applications including renewable energies to stabilize the frequency of the main grid and can be used to supply loads in short-term failures, increasing electrical reliability and stabilizes power fluctuations. Studies have been presented where if the implementation of the flywheel is combined with an active power filter for microgrid application, the power quality is greatly improved since the filter is used to filter the harmonic distortion and the flywheel to stabilize system power and provide an uninterruptible power supply for short-term failures. Now, by combining the flywheel with a traditional battery, needs for large capacity and fast response could be met simultaneously.

3.3 Compressed air

ESS based on compressed air is one of the most promising technology to address multiple problems derived from the high penetration of RES in microgrids due to their characteristics such as less restriction in their construction, high efficiency and respect for the environment. Compressed air is another method of storing energy so that it can be used at some other time, for example during periods of high demand. For this purpose, a turbine is used to expand the compressed gas, which can be transformed into mechanical energy [62]. During the period of low power demand, the excess power drives a reversible motor or generator unit, which in turn operates a set of compressors to inject air into the storage unit. This unit is shaped like an underground cavern. However, during low power generation to meet load demand, the stored compressed air is released and then heated by a heat source. The energy from the compressed air is then transferred to the turbine. It is here that a recovery unit is used to recycle residual thermal energy, further reducing fuel consumption and efficiency. Compressed air can be built from small to large scale. Nevertheless, it is suitable for a large-scale unit involving grid applications such as load shifting, peak-hour demand drop, and voltage and frequency control. The response time of

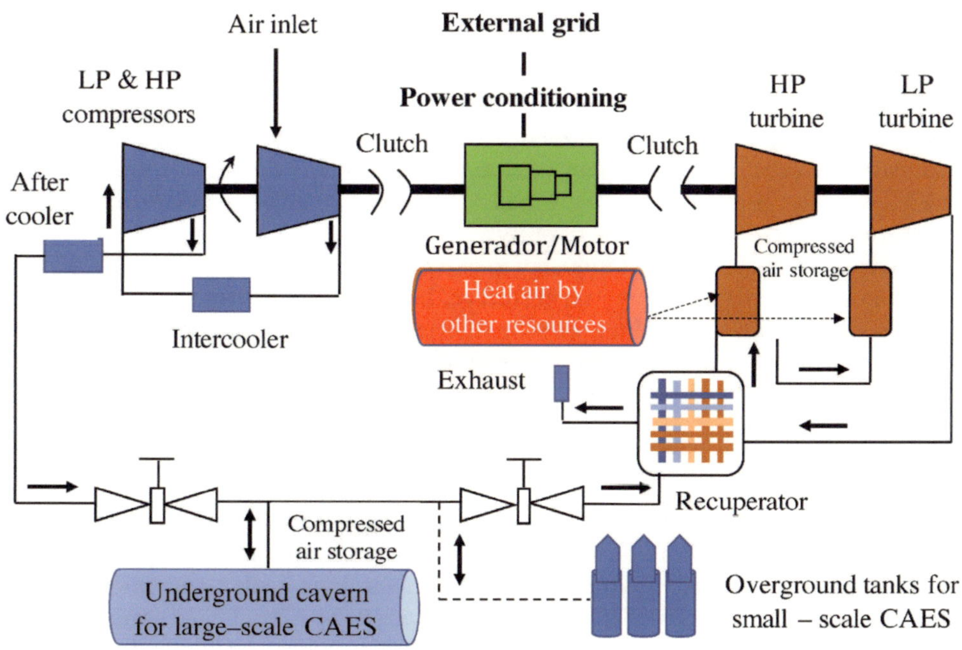

Figure 7.
Simplified diagram of compressed aire storage system.

this system is often high and can smooth energy production in both onshore and offshore wind plants. **Figure 7** illustrates the simplified schematic of a CAES plant.

There are many challenges in implementing this system on a large scale. One of them is the adequate selection of the geographical location with natural underground caverns [63]. For microgrid applications, it has been analyzed that this technology type improves the flexibility and load displacement of the distribution grid and the microgrid itself. In off–peak hours it can be used to supply power to the loads at peak–hours to achieve the economic benefit of the microgrid.

3.4 Fuel cells

Fuel cells as a promising energy source have once again attracted the attention of academia and industry since the beginning of the 21st century because this system type is suitable for the generation of electricity free of toxic emissions and applicable in DG as it has a high energy density by weight and low energy density by volume. In terms of environmental impact, this system type is desirable, leading governments around the world to improve the prospects for the hydrogen economy [64]. The cost per unit of electricity generation for this system has decreased given the raw material resources available, for which there are three types of electrolysis technology: alkaline, polymeric electrolyte membrane and high temperature solid oxide electrolysis [65]. Among these alternatives, alkaline electrolysis turns out to be the most suitable due to its technological maturity and low cost ($525/kW).

The fuel cells integration for microgrid applications has proven to be a promising solution, as it can provide reliable, efficient, clean and quiet energy. In general, according to the role of the fuel cell, four emerging markets can be classified for microgrid applications: primary energy, backup energy, combined heat-energy, and fuel cell vehicles. This integration has several advantages such as economic benefits, prominent energy efficiency, environmental benefits, modularity, improved reliability and power quality. With regard to the latter, in the case of insufficient electricity supply, hydrogen is converted into electrical energy by the fuel cell. That is, the fuel cell can improve the power quality aspects in microgrids and enhance local reliability by balancing power demand and supply, minimizing power fluctuations induced by RES when combined with the electrolyzer to storing and reusing excess energy in the form of hydrogen. For this, the application of hydrogen-based energy storage in a low-voltage microgrid has been studied, achieving good results, where it is proposed that hydrogen cells may have a significant potential to help microgrid in a way effective if a wide range of RES are used. Finally, this system may be economically viable for the mitigation of daily load variability at the site, therefore, additional efforts are needed from academia and industry to explore the multiple uses of hydrogen in a microgrid context such as long-term storage, hydrogen vehicle fuel production, or in combination with the production of synthetic gases. Also, this system is proposed for load displacement applications, however, this technique is expensive and its efficiency is the most critical criterion for developing this techno-logy [66].

3.5 Supercapacitor

ESS based in supercapacitors is one of the best options for microgrid applications due to their high short-term storage capacity, wide operating temperature range, cost-effectiveness, environmental advantages, long cycle times (more than 1×10^5 cycles) and its high efficiency ranging from 84 to 97%. This technology type is also used to energy manage in the microgrid, that is, when the load in the microgrid is light or when the energy supply is ample, the supercapacitors will store energy and when the energy of the microgrid is scarce or when there is some failure in the main grid then the supercapacitors will supply power. Furthermore, they are capable of compensating for power fluctuations derived from load transients and, therefore, can improve power quality as well as extend the useful life of distributed generators [67]. This system type is so versatile that its applications are very varied, in fact, they have great application in the communications area and aeronautics, since they present a rapid response in load leveling and power balance [68, 69]. They also have application in railways, where an efficiency of 55.5% is recorded [70]. However, this system presents several challenges such as a high daily discharge rate of approximately 5–40% and the capital cost is also high, above $6000/kWh. To overcome these challenges, multilayer supercapacitors are proposed, consisting of materials such as carbon, graphene or paper [71] or ultra-small silicon nanoparticles based on polyaniline electrodes [72].

Nowadays, people pay more and more attention to the power quality problem. On the one hand, the microgrid must meet the quality requirements of the load power supply and ensure minimal frequency fluctuation, voltage amplitude and waveform distortion. On the other hand, the main grid establishes strict requirements, such as the power factor limit, the current harmonic distortion rate and the maximum power to incorporate the microgrid as a whole with the main grid. That is why, through the inverter control unit, supercapacitors can be adjusted to provide active and reactive power to users, in order to improve power quality. STATCOM, in conjunction with supercapacitors, is also used to improve the power quality in microgrid. Finally, for uncontrollable micro-sources such as wind and solar, fluctuations caused by the power output of generators will decrease by improving power quality. This union of the STATCOM with supercapacitors solves the dynamic power quality problems, such as the voltage drops, the harmonic currents and the instantaneous voltage interruption by the combination of the sources with ESS. **Figure 8** illustrates the principal structure of a supercapacitor.

3.6 Superconducting magnetic energy storage (SMES)

This technology type works based on the electrodynamics principle [73] where energy is stored in a magnetic field created by the DC flow in a superconducting

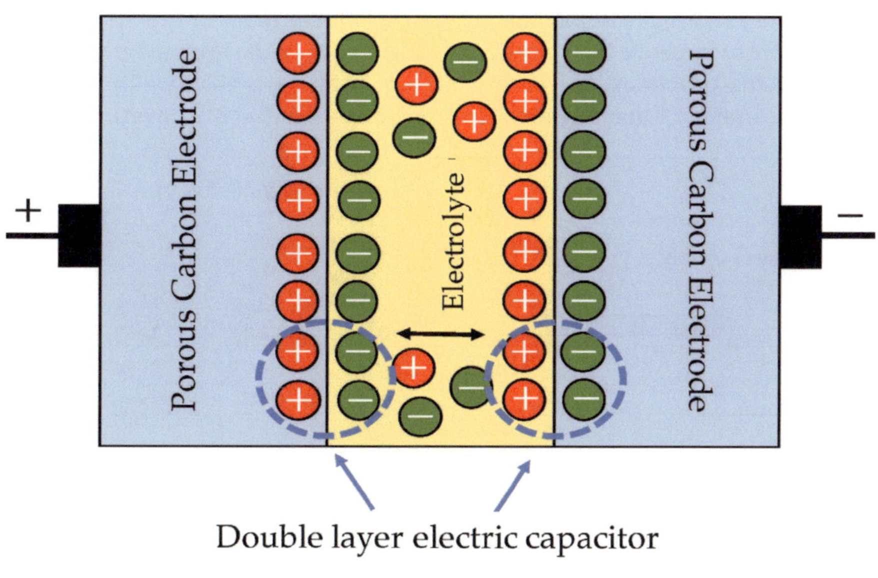

Figure 8.
Schematic view of supercapacitor.

magnetic through an AC–DC converter (charge mode). However, the stored energy can be delivered back to the electrical grid using a DC–AC converter (discharge mode). This ESS has the drawback of having ohmic losses which generate heat in the system and, therefore, cause thermal instability in the superconducting magnetic [74]. This type of storage is classified into two types: high-temperature superconducting magnetic (HTSM) that operate at approximately 70° K and low–temperature superconducting magnetic (LTSM) that operate at approximately 7° K. **Figure 9** shows the basic diagram of the SMES system.

LTSM is a system that presents greater technological maturity compared to the HTSM since it provides a rapid response to the charge and discharge cycles in a few milliseconds. Among its most important characteristics are: high energy density (4 kW/l) and high efficiency, around 95–98% with a long service life of approximately 30 years. This system is available on the market in a wide range of powers ranging from 0.1–10 MW. With the system advancement, the capacity of this system is expected to increase to 100 MWh in the next 10 years. However, due to the complexity of the cooling system, the material of coil manufacture and the superconducting cables, the cost of installing is high, around $10,000/kWh [50], therefore only they are used for short-term energy storage [75]. Finally, the SMES are highly applicable in microgrids due to their flexible capacity to exchange active and reactive power and thus improve the power quality, the power factor and stabilize the frequency. They also play an important role in the RES integration such as wind generators by controlling the output power of the wind farm and improving the stability of the electrical system. Actual research on this system type is based on reducing the cost of coils and cooling systems to result in an attractive and competitive system for users.

Table 1 summarizes the most important general characteristics of the ESS for microgrid applications described in this section and **Table 2** summarizes the most important problems of the power quality generated in microgrids and the ESS that are used for mitigating those problems.

3.7 Hybrid ESS

Hybrid Energy Storage Systems (HESS) refers to the integration of two or more ESS in order to achieve greater advantages and characteristics of high power and energy in order to improve the stability and reliability of the system by minimizing the power quality problems [85]. A proposal of the above is observed in **Table 2**. HESS control strategy is usually more complicated than that of conventional ESS

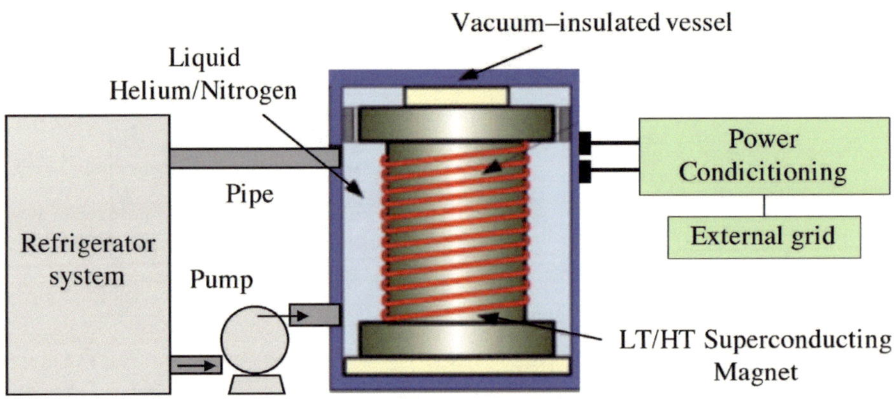

Figure 9.
Principal diagram of SMES system.

ESS	Energy Density (WH/L)	Power Density (WH/L)	Nominal Power (MW)	Life time (years)	Discharge efficiency	Response time	Storage duration	Discharge time at rated power	Technological maturity
Batteries	200–500 [56, 76]	1500–10,000 [76]	0–100 [56, 76]	5–16 [76, 77]	85% [56, 77]	Milisec. [77]	Min–Days [76]	Min–Hrs. [76, 78]	Mature
Flywheels	20–80 [76]	1000–2000 [76]	0.1–20 [76]	15–20 [76, 79]	90–93% [61]	Seconds [80]	Sec–Min. [76, 81]	Sec–Min. [76]	Market
Compressed air	2–6 [62, 76]	30–60 [62, 76]	300–1000 [63, 76]	20–40 [62, 76]	70–80% [82]	Minutes [62]	Hrs–Days [76]	1–20 Hrs [76, 82]	Market
Fuel cells	500–3000 [64, 76]	500 [76]	0–58.8 [65]	5–20 [66, 76]	59% [83]	Seconds [83]	Hrs–Days [76]	Sec–24 Hrs [76]	Developing
Supercapacitors	10–30 [67, 76]	100,000 [76]	0–0.3 [67, 76, 84]	10–30 [76, 85]	95–98% [86]	Milisec. [86]	Sec–Hrs [76, 87]	Milisec–1 Hr [76, 88]	Developing
SMES	0.2–6 [76]	1000–4000 [76]	0.1–10 [76]	20–30 [76]	95% [89]	Milisec. [90]	Min–Hrs [76]	Milisec–30 Min [76, 91]	Developing

Table 1.
Actual characteristics of ESS with electrical microgrids applications.

ESS	Power quality problems	Ref.
Batteries	Harmonic distortion	[92]
	Reactive currents	[82]
	Voltage sags	[93]
Flywheels	Voltage sags	[82]
Compressed air	Voltage fluctuations	[90]
	Frequency variation	[94]
Fuel cells	Voltage unbalance	[95, 96]
Supercapacitors	Voltage sags	[87]
	Harmonic distortions	[88]
	Voltage interruptions	[88]
SMES	Voltage sags	[97]
	Resonances	[97]

Table 2.
ESS implementation for microgrids applications for the power quality problems mitigation.

High energy supplier storage device	High power supplier storage device	Ref.
Batteries	Supercapacitors	[98, 99]
	SMES	[100]
	Flywheels	[101, 102]
Compressed air	Supercapacitors	[103]
	SMES	[92]
	Flywheels	[104]
	Batteries	[93]
Fuel cells	Supercapacitors	[105–107]
	SMES	[108]
	Flywheels	[109]
	Batteries	[110]

Table 3.
Possible configuration of the Hybrid ESS.

since they consider characteristics such as: charge and discharge, response time, energy distribution, life cycle and efficiency. HESS has been a subject in which several researchers from around the world have been involved, which use various storage techniques whose proposals are listed in **Table 3**. **Figure 10** shows hybrid ESS topology for microgrid applications.

In summary, the HESS for microgrid applications showed a better performance in stabilizing the frequency compared to the conventional ESS, for example, the battery life cycle improves due to obtaining protection against high charge and discharge cycles, frequency and against very high currents. Combining the batteries with some other technology extends their useful life from 5.7 to 9.2 years.

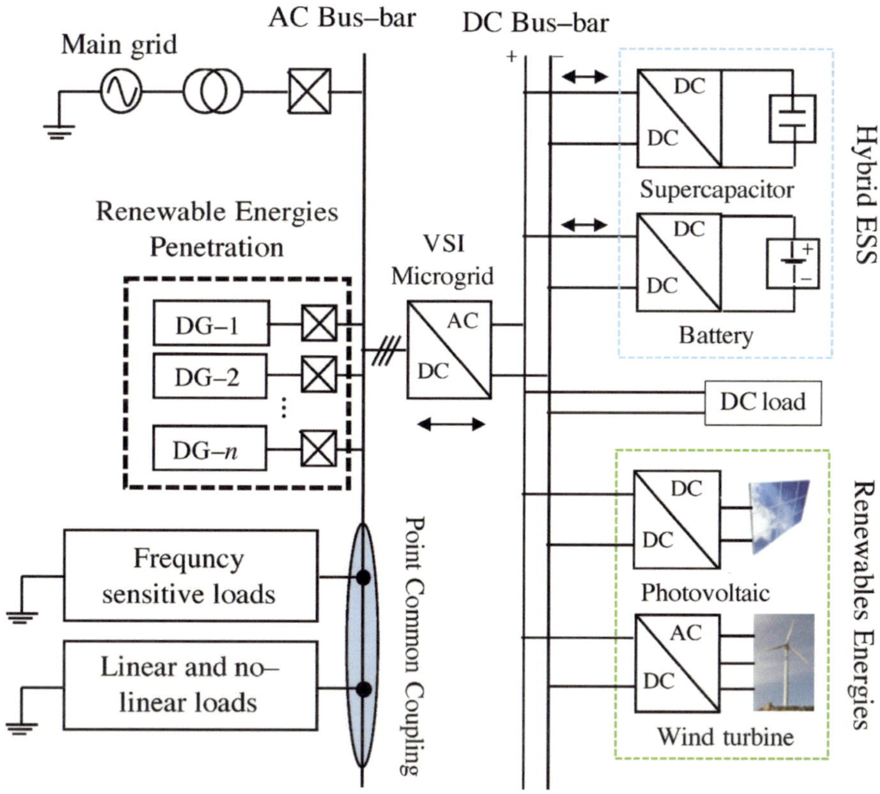

Figure 10.
Hybrid ESS topology for microgrid applications.

4. Overview of microgrid power quality

The term power quality is typically used for a wide range of electromagnetic events generated in electrical power systems. Therefore, many researchers have focused their studies on this topic and in recent years have published important findings about the problems of power quality when connecting microgrids to the electrical grids [111]. Power quality problems have recently become important given the need for reliable power to meet customer needs and the presence and extensive use of different types of electronic and electrical appliances in the commercial and industrial sectors. **Table 4** shows the PQ problems introduced in different DG units.

Power quality is a major concern in small-scale island or monovalent microgrids due to the presence of both non-linear and unbalanced loads, which make up a larger proportion of the total microgrid load. This situation creates voltage problems such as distortion, fluctuation, and sags/swells in a relatively weak system [112]. In a microgrid operating in island mode, disturbances such as distortion or voltage unbalance are more likely to occur due to very high impedance levels as well as load distribution compared to microgrids operating in grid-connected mode. In this mode of operation, the most frequent problems are disturbances and unbalanced voltages from the grid [113]. The voltage generated by sources such as wind, solar energy and fuel cells is intermittent and therefore these sources cannot be directly connected to the grid. **Table 3** shows the power quality problems introduced in different DG units. Power quality problems are analyzed based on the development of standards which define acceptable levels of distortions and deviations in various electrical quantities, such as current, voltage, and power factor.

Power Quality Issues	Solar	Wind	Small–hydro	Diesel
Voltage (sag/swell)	×	✓	✓	✓
(Over/under) voltage	×	✓	×	✓
Voltage unbalance	✓	×	×	×
Voltage transient	×	✓	×	×
Voltage harmonics	✓	✓	✓	×
Flicker	✓	✓	×	✓
Current harmonics	✓	✓	✓	×
Interruption	✓	✓	×	×

Table 4.
Power quality issues related to generation units of microgrid.

4.1 Sag/swell issues in microgrid applications

Voltage sag represent one of the most serious power quality challenges which are mainly caused by failures and lead to power sector instability, interruption in the operation of sensitive electronic devices, which is typical in microgrids that consist of DES. On the other hand, voltage swell, whose behavior is the opposite of voltage sags, is another serious power quality problem; however, it rarely occurs [114]. As the integration of DES in microgrids increases, many standards and grid codes impose new regulations, such as the ability to withstand voltage sags (LVRT) and voltage swell (HVRT). These regulations require that the microgrids disconnect from the grids in case the voltage sags or swell has a specific duration [115]. In the case of voltage sags, the German standard dictates that the microgrids must remain connected and withstand the event by providing reactive power even if the voltage sags to 0% of its nominal value for 0.15 s; otherwise, disconnection is mandatory. Otherwise, for voltage swell, the German standard dictates that the microgrid must remain connected even if the voltage swell to 120% of its nominal value for 0.1 s; otherwise, disconnection is mandatory [116]. The voltage level and duration of both the voltage sag and swell differ from one grid code to another.

4.2 Harmonics

Non-linear loads, electronic inverters, computer controllers, and variable speed motors that generate harmonics are applicable for microgrids. Most electrical system handle harmonics down to a specific amount; however, once the number of harmonics is large, it will cause communication failures, excessive line losses, overheating and tripping of the circuit breaker [117]. For this reason, many studies have been carried out in low voltage systems to analyze the power quality with respect to harmonic distortion problem. Considering that an microgrid is a low-voltage grid, then harmonic distortion as a severe power quality problem is an important problem for this type of system, and it should be investigated and addressed [118]. The sources that make up the microgrid consist mainly of RES with a power electronics device that produces harmonics in the system. Therefore, microgrids must reduce the emission of harmonics in accordance with what is dictated by current standards and codes [119].

As the specifications for the integration of microgrids into the main grids progress, various criteria about harmonic distortion is also implemented to ensure that the voltage and current are compatible with the grids as much as possible. Therefore, some requirements have been imposed on the limits of total and individual harmonic distortion (THD) for microgrids connected to the main electrical grid [120–122]. For current THD (THD_i), all requirements, standards and grid codes are similar, that is, it should be less than 5%. UK standards (EREC G83) are more stringent and require THD_i <3% [123]. Regarding the voltage THD (THD_v), the literature indicates that most countries follow the IEEE or IEC standards [124], in which the THD_v should not exceed 5% in a microgrid.

4.3 Voltage unbalance and fluctuation

Voltage unbalance is the most frequently occurring power quality phenomenon. The voltage unbalance factor (VUF), which is the ratio between the positive and negative sequence of voltage components, is used to measure the degree of unbalance in the system [125]. Voltage unbalance can have adverse effects on microgrids power electronics as well as power system devices. Power systems will suffer a greater number of losses and will be less stable in unbalanced conditions; therefore, it is essential to have a balanced system, especially with the diversity of sources that make up the microgrid [126]. In addition to the above, certain criteria have been established in the grid codes and standards to guarantee a stable and balanced integration of the microgrids to the main grid to limit the VUF. For example, IEEE Std [127] does not allow the VUF to exceed 3%. The IEC standard requires that all distribution generators keep the VUF below 2% [121]. The requirements of China and Germany state that the VUF should not exceed 2% [128, 129]. The Canadian Standards Association (CAN/CSA–C61000–2–2) established 2% as the maximum allowable VUF limit; for the case of unbalanced loads, 3% is allowed [130]. Generally, global standards indicate that the acceptable limit of VUF should be between 1% and 2% [131].

Fluctuations in microgrids are known as slow switch voltage variations or stable operations. Typically, voltage fluctuation in microgrids occurs due to changes in solar irradiation, wind speed, battery charge/discharge, and charge variations [132]. An essential detail to mention is that voltage fluctuations can be caused by sources whose output power changes widely over time.

5. Power quality mitigation devices, methods, and control strategies in microgrid applications

Power quality plays an increasingly important role in both energy supply and demand. With the participation of private companies in the distribution systems, it is expected that the power quality will be the deciding factor for consumers. Due to the increasing application of switching devices, power quality is likely to deteriorate. For this reason, this situation has drawn the attention of researchers to identify and suggest mitigation strategies for power quality problems to improve the microgrid performance.

5.1 Electronic controllers

Numerous studies reveal that advanced control technologies have been adopted to reduce the negative effects caused by the main grid connected to DG, undeniably improving aspects of power quality. Photovoltaic systems with voltage monitoring controllers have been established based on various control theories, achieving a significant improvement in the transition process during the connection of the photovoltaic systems that make up the microgrid [133]. Also, in the wind energy area, fuzzy logic-based controllers have been designed to control the inverter and PMSG (permanent magnet synchronous generator) operation, however, this could cause oscillations near the operating point [134]. The smart solution concept has been introduced to mitigate the grid–side converter voltage ripple and improve certain aspects of power quality as well as the efficiency of the grid connected to a photovoltaic system for microgrid application [135]. Finally, the experimental proposal of a magnetic flux control applied to a variable reactor integrated to a power quality controller has generated good results which validate the controller's capacity by mitigating a large percentage of harmonic penetration.

5.2 Dynamic voltage restorers (DVR)

A DVR is used to mitigate power quality problems in microgrids, mainly voltage sags and swell, thus improving the power quality of microgrids containing PV and batteries [136]. However, there are still some limitations in terms of LVRT. Therefore, in [137] they use an optimization technique to improve the performance of the DVR and thus solve the problem of voltage sags in microgrids using fuzzy logic. The effectiveness of this method reduced the VUF to less than 1%, while the current and voltage THD was reduced to less than 5%, as indicated in the current grid codes. Overall, DVR is one of the best devices to mitigate power quality issues when the microgrid is connected to the main grid. Finally, the equivalent circuit of the DVR is shown in **Figure 11**.

5.3 STATCOM and SVC

STATCOM and SVC are other devices used to solve power quality problems. These two devices, shown in **Figure 12**, are flexible AC transmission system devices and have been widely used in recent years to solve many power quality problems mainly due to RES integration, such as LVRT, to overcome voltage swell in PV systems [138] and wind systems [139]. The authors compared the efficacy of SVC and STATCOM in addressing voltage sag problems and found that STATCOM contributes more to the transient margin compared to SVC. In [140], STATCOM was used

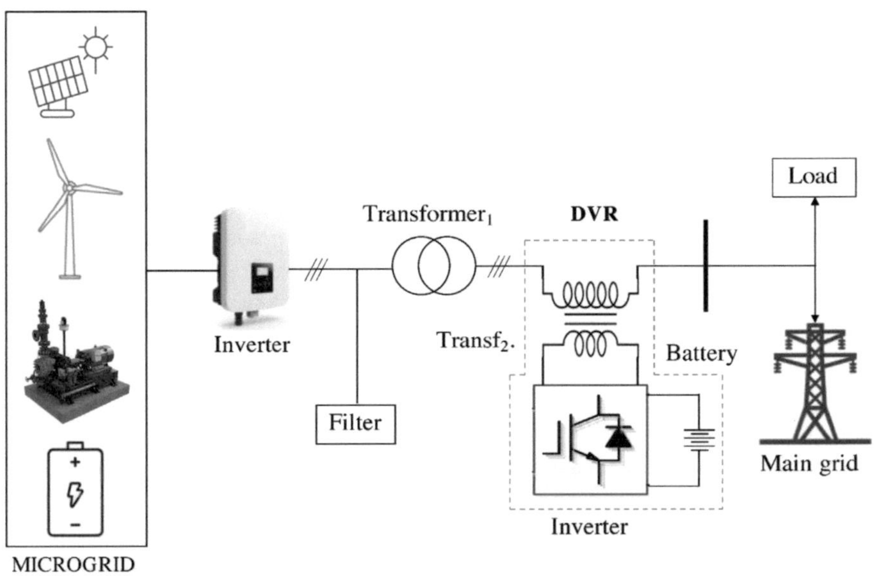

Figure 11.
DVR – integrated microgrid system to mitigate power quality issues.

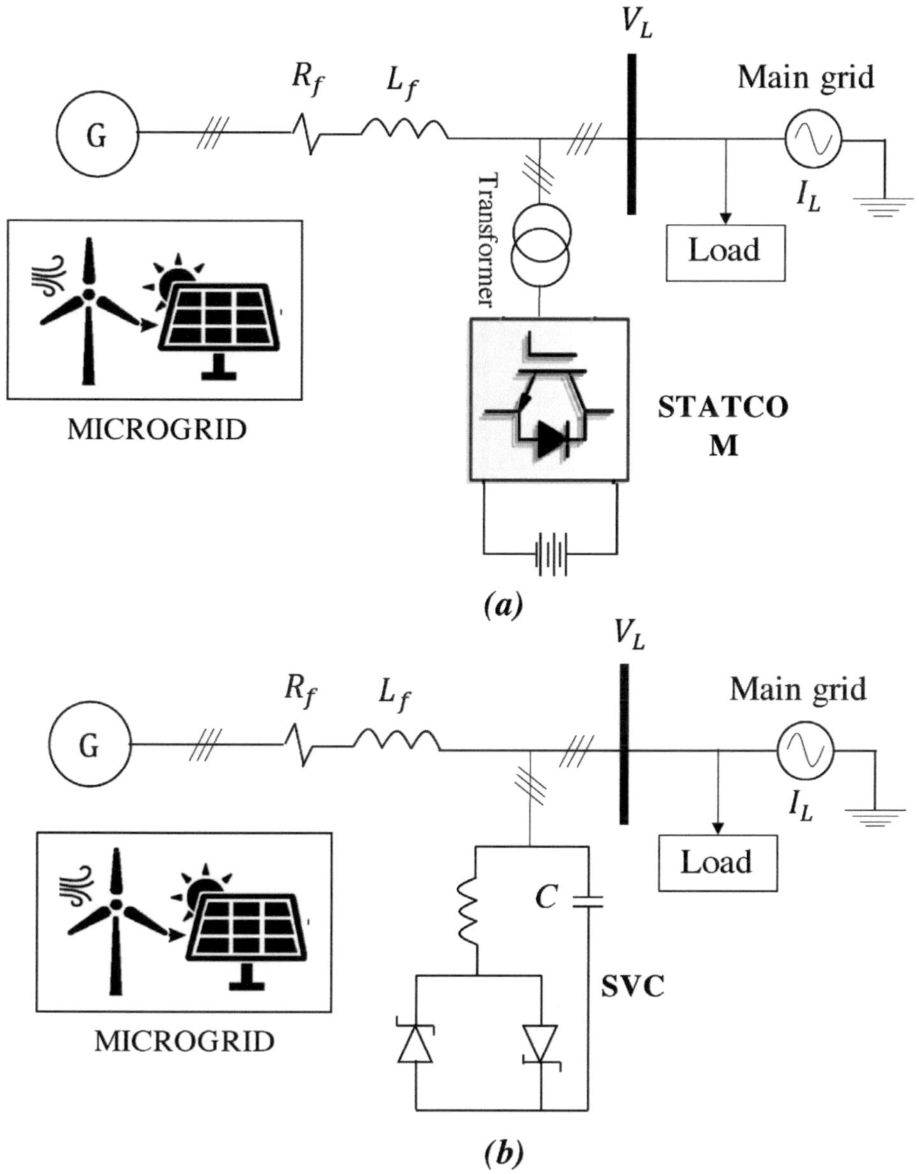

Figure 12.
Typical configuration of: (a) STATCOM and (b) SVC used for power quality mitigation in microgrid systems.

to mitigate voltage fluctuations at high penetration DER for microgrid applications. Furthermore, STATCOM was used to mitigate voltage fluctuation and compensate for reactive power in microgrids in [141].

Another study demonstrated the ability of STATCOMs to reduce power fluctuation in microgrids and increase voltage regulation and power factor of the system. Regarding the mitigation of harmonics and THD in microgrids that use numerous RES such as wind turbines, diesel generators, fuel cells, microturbines and photovoltaic systems, the STATCOM reduced the harmonics according to the IEEE 1547 [142] standards. From the previous studies, it is concluded that STATCOM has a high ability to mitigate voltage fluctuation and improve the voltage profile in microgrids while mitigating voltage sags/swell to a lesser extent. The SVC was used in microgrids to improve the power quality of the grid and to increase the efficiency of the system during voltage sag. The authors in [143] used it for the same purpose in a microgrid operating in island mode where it showed good performance. However, during severe brownout events, the SVC performs worse than the DVR and STATCOM.

5.4 FACTS

A Flexible AC Transmission System (FACTS) is a system composed of static equipment used for the transmission of electrical energy in AC. This device is intended to improve the controllability and increase the power transfer capacity of the main grid. Relevant studies incorporate a distribution static compensator (D–STATCOM), which injects a reactive component to provide rapid voltage regulation at the load terminal while maintaining an almost unity power factor (PF=1) [144]. IR (Intelligent Detection & Reconnection Technique), with a UPQC (Unified Power Quality Conditioner), is used for secondary control of the direct current link integrated to ESS. In addition, this technique compensates for voltage interruptions, reactive power, voltage drops, and harmonic distortion [145].

5.5 Unified power quality conditioner (UPQC)

The UPQC is the complete hybrid filter configuration and is identified as a multi-functional power conditioner used to compensate for different voltage disturbances, correct voltage fluctuations and prevent the entry of harmonic currents in the electrical grid. Originally, it was designed to mitigate disturbances that affect the performance of sensitive and/or critical loads and thus improve the power quality of the electrical system. The UPQC is a combination of serial and parallel controllers connected by a common DC bus, as shown in **Figure 13**. The parallel controller can generate or absorb reactive power at the point of connection. However, the serial controller is linked with the microgrid to control the parameters of line [146]. In [147], the fuzzy logic technique was implemented in a UPQC to minimize harmonic voltages and harmonic currents. The results show that the overall harmonic distortion was reduced from 8.93% to 3.34%. In [148] the design

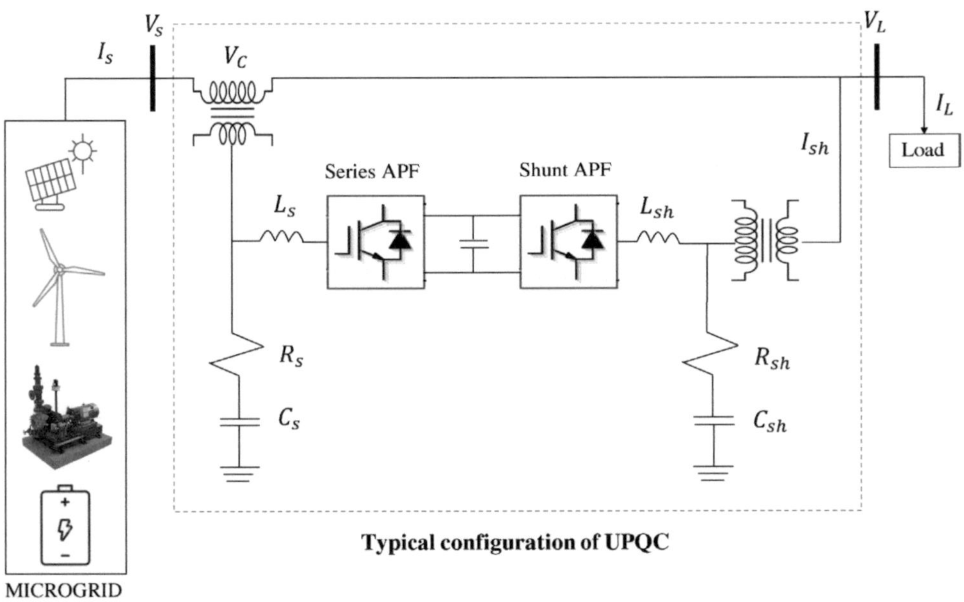

Figure 13.
Typical configuration of: (a) STATCOM and (b) SVC used for power quality.

Factors	DSTATCOM	SVC	DVR	UPQC
Rating	Low rating	Low	High rating	Higher ratings are available
Speed of operation	Less than DVR	Less than DVR	Fast	Faster
Compensation method	Shunt compensation	Shunt	Series compensation	Both series and shunt
Active/reactive power	Reactive	Reactive	Active/reactive	Both
Harmonics	Less	Less	Much less	Least
Problems addressed	Sag, swell	Sag, swell	Sag, harmonics, fluctuation, swell	Swell, sag, harmonics, transient, unbalance and flicker
Cost	Normal	Average	High	Higher
Complexity	High	High	High	Higher

Table 5.
Comparison of various custom power devices.

of a suitable UPQC implemented in a microgrid is proposed to improve the harmonic distortion. The results show that the measured voltage sag occurs from 0.2 to 0.3 s with a THD of 2.69%. The analysis of the harmonic spectrum of the current without considering the UPQC shows a THD of 33.26%. Using UPQC you get a current THD of 3.11%, which meets IEEE 519–1992 standards of less than 5%. The UPQC is used in to mitigate the sags/swell in a microgrid consisting of a hybrid PV/wind system by injection or absorption of reactive current. Mitigation of voltage sag and reduction of THD using the UPQC device using an adaptive neuro-fuzzy inference system (ANFIS). In [149], a UPQC is designed to power quality improve, and its performance was evaluated for various non-linear loads. Results show that UPQC reduced THD_i and THD_v when ANN control techniques were used to improve overall performance. THD is reduced from 12.6% to 3.7% and from 7.34% to 3.7% for voltage and current, respectively. Although UPQC is widely used to mitigate harmonics, the authors in [150] introduced UPQC to mitigate voltage unbalance in the microgrid connected to the grid. The results illustrate that the UPQC can detect the incidence of voltage unbalance and reduce the VUF to less than 2%, as established by grid codes.

Finally, based on the literature described, **Table 5** illustrates a comparison between the most popular devices used to mitigate power quality problems in microgrids. The comparison was made in terms of cost, maturity and performance. Overall, the DVR is superior to SVC and STATCOM in terms of voltage sags and/or swell, voltage fluctuations and unbalances, while the UPQC offers the best protection for sensitive loads from low quality sources.

5.6 Mitigation techniques

To improve power quality and, consequently, microgrid performance, harmonic mitigation techniques are implemented which can be: passive techniques, multi-pulse rectifier techniques, and active harmonic cancellation techniques. For low-power industrial applications, traditional harmonic mitigation techniques such as AC–DC inductors are used due to their low cost, reliability, and simplicity [151], as well as passive harmonic mitigation techniques to analyze grid disturbances. In the latter, harmonic mitigation depends, to a great extent, on the grid configuration [152].

Additionally, methods have been implemented for estimating the positive sequence phase angle based on the Discrete Fourier Transform [153] where the transient response of a single cycle that is immune to harmonics, electromagnetic noise, unbalance voltage, and frequency variations in microgrid applications. The implementation of "shunt" type active power filters is another way to improve the power quality of a microgrid at distribution level using intelligent learning algorithms [154]. Another way to counteract power quality problems due to the RES incorporation is through active power filter control methods where the inverter injects the power generated by the renewable source into the main grid (it works as an active power filter and injects power into the main grid). Basically, the inverter operates in two modes, (a) mode 1: it injects power generated by RES, improving the power quality, and (b) mode 2: no power is generated and acts as a "shunt" type active power filter [155].

6. Issues and challenges of the ESS for microgrid applications

Carbon emission drives the world to replace conventional power generation with as much renewable generation as possible. However, when integrating the energy generated by RES to the grid, problems are generated in the power quality, for which new techniques must be evaluated to mitigate these problems and thus improve the microgrids performance that contain RES. The power quality concept is not strictly defined yet for microgrid application. Therefore, the responsibility for maintaining a good electrical environment rests with the dealer, the manufacturer and the user. As previously mentioned, DG generates power quality problems: power flow variation which causes voltage and frequency deviations, voltage and current unbalance, power factor poor, harmonic distortion, voltage flickers as well as voltage drops, among others. Faced with these problems, ESS play a key role when meeting different needs of this kind. Depending on the application field, there are three general cases: (i) storage for power quality improvement, (ii) emergency storage and (iii) storage for grid management. In the first case, the power quality can be improved with systems that give up the accumulated energy in the shortest possible time (seconds), as is the case with supercapacitors that with up to 95% efficiency and barely 5% of daily self-discharge losses, they are capable of storing an unusually high energy density. On the other hand, emergency storage is designed to provide energy, activating in a matter of minutes and remaining in operation to ensure continuity of supply. This is the case with flywheels coupled to generators, which allow short-term energy storage with efficiencies of up to 85%. However, being mechanical systems, this figure drops by up to 40% after just one day of storage due to friction losses. The kinetic energy they store is designed to be returned through generators to support sudden changes in electricity demand. Finally, storage for grid management, which are systems that allow the prolonged storage of large amounts of energy and that are capable of gradually transferring it to support the management of the grid itself. This is the type of energy storage that allows renewable energy to be stored for later use in a manageable way. As can be seen, there are still niches of opportunity for the research field regarding the development of ESSs and their microgrid application to meet the requirements of international standards in the power quality area. The most important challenges faced by ESSs and their microgrid application are listed below.

6.1 Suppression of power fluctuations

RES, such as wind and solar, are often unstable energy sources. The power production of the wind turbines and solar panels is intermittent due to climatic variations such as clouds on the photovoltaic panels or the wake effect of the wind turbine as well as the shadow effect of the tower. Knowing that microgrid is not as strong as the grid, power fluctuations cause power quality problems, specifically notable variations in grid frequency and voltage, making microgrid unstable.

6.2 LVRT capability

When some incidents occur in the microgrid or in the main grid, the voltage can drop suddenly and cause some GD sources to disconnect, ceasing to produce power. If a certain GD source in the microgrid trips due to voltage dips, this can cause other GD sources to disconnect from the main grid and cause a cascading power outage. Therefore, GD sources, such as wind turbines or solar panels, are required to remain connected to the grid during the presence of voltage sags, that is, they have the ability to operate in low-voltage conditions. According to [156], by integrating the ESS into the microgrid, the capacity of wind generators to operate in voltage dips can be improved.

6.3 Spinning reserve

Power generation is highly dependent on weather conditions, which are constantly changing. As a result, power shortages can occur more frequently into the microgrid. Therefore, the spinning reserve issue is very important for the microgrid operation. With the ESS implementation, most of the spinning reserve requirements in the generators for microgrid applications will be able to be met, that is, with the help of the ESS, the generators will be able to operate closer to their nominal value. Also, ESS can react faster than many generators, so power shortages can be quickly recovered. The most important problem with spinning reserve technology is determining how much power to reserve. Both the reliability and economics of the microgrid must be considered when deciding the amount of spinning reserve. This will undoubtedly improve the microgrid performance and power quality aspects [157].

7. Conclusion

Actually, ESS and the availability of mitigation methods are an alternative solution for the potential use of RES for microgrid applications. Many researchers are involved in the development of ESSs and their microgrid applications to manage the energy balance by storing it during off-peak hours at reduced cost. Therefore, an optimal ESS model is the key to a successful storage future. However, efficient development of ESSs for microgrid applications is a challenge in power quality terms. Numerous studies and reviews about ESS are limited to analyzing the types, characteristics, configurations as well as the operational advantages and disadvantages, but very little is addressed to the issue of improving the power quality into microgrids through the ESS. Therefore, the key contribution of this study has been the exhaustive analysis of the ESS actual state for microgrid applications as well as the issues and challenges they face in meeting power quality standards. This review proposes some technical and operational suggestions:

- Advanced research is required to improve the capabilities of ESSs for microgrid applications in terms of materials, size, cost and efficiency considering the adequate functionality of the system and its acceptance in the market.

- An advanced power electronics system in conjunction with ESSs could help overcome switching challenges and power quality issues into microgrids, addressing issues such as overheating, harmonic distortion, and charge–discharge for efficient operation of the system.

- Development of appropriate techniques for the ESS optimal sizing and thus ensure efficient operation in terms of: energy arbitrage, energy backup, energy demand at peak hours and voltage support.

- Advanced research is required on the integration of ESSs for microgrid applications, addressing the issue of complexity in synchronization, improving the performance of integration or operation in "island" mode.

These suggestions would be notable contributions towards the maturity of ESSs, which are expected to dominate the electricity market in the future. In addition, from this review, some important and specific recommendations relevant to power quality mitigation issues and techniques are summarized below for further improvement:

- Further studies should consider more RES to conform to microgrids, such as hydropower, biomass and geothermal, along with non-RES such as diesel generator to show the effects of a wide variety of sources on energy quality.

- Devices such as DVR and UPQC must use a fast and accurate method to detect power quality problems in microgrids.

- In the future, generalized validation and benchmarking methods can be applied for mitigation of power quality in microgrids using optimization methods that take into account uncertain climatic conditions.

- International system operators should adopt a single or constant limit for each integration requirement to reduce differences between current technical requirements and thus harmonize power quality requirements in microgrids.

The above recommendations may be the most important contributions towards improving the power quality in microgrids, especially with renewable generation sources, which are expected to dominate the energy market in the near future. Future studies based on the results of this review may also help to address current drawbacks of microgrids in developing new standards and preventing new power quality problems.

Conflict of interest

The authors declare no conflict of interest.

Author details

Emmanuel Hernández Mayoral[1*], Efraín Dueñas Reyes[2], Reynaldo Iracheta Cortez[3], Carlos J. Martínez Hernández[2], Carlos D. Aguilar Gómez[4], Christian R. Jiménez Román[5], Juan D. Rodríguez Romero[5], Omar Rodríguez Rivera[5], Edwin F. Mendoza Santos[6], Wilder Durante Gómez[7] and José I. Barreto Muñoz[2]

1 CONACyT Attachment at Institute of Renewable Energies, Temixco, Mor., Mexico

2 Isthmus University, Cd Univers., Tehuantepec, Oaxaca, Mexico

3 CONACyT attachment at Isthmus University, Cd Univers., Tehuantepec, Oaxaca, Mexico

4 Polytechnic University of Chiapas, Suchiapa, Chiapas, Mexico

5 Institute of Renewable Energies, Temixco, Morelos, Mexico

6 Technological Institute of Celaya, Antonio García, Celaya, Guanajuato, Mexico

7 Faculty of Engineering, Mexico City, Mexico

*Address all correspondence to: emhema@ier.unam.mx

References

[1] "EPRI–DOE of Energy Storage for Transmission and Distribution Applications," EPRI, and the U.S. Department of Energy, Palo Alto and Washington, CA and DC, 2003, EPRI–DOE no. 1001834.

[2] S. B. groupe Energie, "Energy Storage Technologies for Wind Power Integration," Tech. Rep., Université Libre de Bruxelles, Faculté des Sciences Appliquées, 2010.

[3] "EPRI–DOE Handbook Supplement of Energy Storage for Grid Connected Wind Generation Applications," EPRI, and the U.S. Department of Energy, Palo Alto and Washington, CA and DC, 2004, EPRI–DOE no. 1008703.

[4] Manz D, Schelenz O, Chandra R, Bose S, de Rooji M, Bebic J. Enhanced Reliability of Photovoltaic Systems with Energy Storage and Controls, Tech. Rep., National Renewable Energy Laboratory, 2008.

[5] Ton D, Peek GH, Hanley C, Boyes J. Solar Energy grid Integration Systems Energy Storage (SEGIS-ES), Sandia Nat. Labs, 2008.

[6] F. Diaz F, Bianchi FD, Sumper A, Gomis O. Control of a Flywheel Energy Storage System for Power Smoothing in Wind Power Plants, IEEE Transaction on Energy Conversion, 2014;**29**:204-214.

[7] Guerrero JM, Loh PC, Lee TL, Chandorkar M. Advanced control architectures for intelligent microgrids Part II: Power quality, energy storage, and AC/DC microgrids, IEEE Trans. Ind. Electron., 2013;**60**:1263-1270.

[8] Liu F, Liu J, Zhang H, Xue D. Stability Issues of Z+Z Type Cascade System in Hybrid Energy Storage System (HESS), IEEE Trans. Power Electron. 2014;**29**: 5846-5859.

[9] Wang P, Xiao J, Setyawan L. Hierarchical Control of Hybrid Energy Storage System in DC Microgrids, IEEE Trans. Ind. Electron. 2015;**99**:1-15.

[10] Han J, Solanki SK, Solanki K. Coordinated predictive control of a wind/battery microgrid system, IEEE J. Emerg. Sel. Top. Power Electron. 2013; **1**:296-305.

[11] Tan X, Li Q, Wang H. Advanced and trends of energy storage technology in Microgrids, Int. J. Electr. Power Energy Syst. 2013;**44**:179-191.

[12] Bhuiyan FA, Yazdani A. Energy storage technologies for grid–connected and off–grid power system applications, 2012 IEEE Electr. Power Energy Conf. EPEC 2012, 2012:303-310.

[13] Katsanevakis M, Stewart RA, Lu J. Aggregated applications and benefits of energy storage systems with application–specific control methods: A review. Renewable Sustainable Energy Review. 2017;**75**:719-741.

[14] Rohit AK, Rangnekar S. An overview of energy storage and its importance in Indian renewable energy sector, Journal of Energy Storage. 2017;**13**: 447-456.

[15] Lasseter RH. MicroGrids. 2002;305-308.

[16] Bragard M, Soltau N, Thomas S, Doncker RW. The balance of renewable sources and user demands in grids: Power electronics for modular battery energy storage systems. IEEE Trans. Power Electron. 2010;**25**:3049-3056.

[17] Levron Y, Shmilovitz D. Power systems optimal peak–shaving applying secondary storage. Electr. Power Syst. Res. 2012;**89**:80-84.

[18] Mohd A, et al., Challenges in integrating distributed Energy storage systems into future smart grid. 2008 IEEE Int. Symp. Ind. Electron. 2008:1627-1632.

[19] Díaz F, Sumper A, Gomis O, Villafáfila A. A review of energy storage technologies for wind power applications. Renew. Sustainable Energy Review. 2012;**16**:2154-2171.

[20] Huff G, et al., DOE/EPRI 2013 electricity storage handbook in collaboration with NRECA, Rep. SAND2013- ..., no. July, p. 340, 2013.

[21] Feltes JW, Grande-Moran C. Black start studies for system restoration. IEEE Power Energy Soc. 2008 Gen. Meet. Convers. Deliv. Electr. Energy 21st Century, PES, pp. 1-8, 2008.

[22] Atwa YM, El-Saadany EF, Salama MMA, Seethapathy R. Optimal Renewable Resources Mix for Distribution System Energy Loss Minimization. IEEE Trans. Power Systems. vol. 2010;**1**:360-370.

[23] Mundackal J, Varghese AC, Sreekala P, Reshmi V. Grid power quality improvement and battery energy storage in wind energy systems. 2013 Annu. Int. Conf. Emerg. Res. Areas 2013 Int. Conf. Microelectron. Commun. Renew. Energy, 1-6, 2013.

[24] Carrasco JM, et al., Power Electronic Systems for the Grid Integration of Renewable Energy Sources: A Survey. IEEE Trans. on Industrial Electronics. 2006;**53**:1002-1016.

[25] Hill CA, Such MC, Chen D, Gonzalez J, Grady WM. Battery Energy Storage for Enabling Integration of Distributed Solar Power Generation. IEEE Trans. Smart Grid. 2012;**3**:850-857.

[26] Subburaj AS, Pushpakaran BN, Bayne SB. Overview of grid connected renewable energy based battery projects in USA. Renew. Sustain. Energy Reviews. 2015; **45**:219-234.

[27] Leou R. Electrical Power and Energy Systems An economic analysis model for the energy storage system applied to a distribution substation. Int. J. Electr. Power Energy Syst. 2012;**34**: 132-137.

[28] Saez A. Analysis and Comparison of Battery Energy Storage Technologies for Grid Applications. IEEE Grenoble Conference, June 2013.

[29] US State Department, "US Climate Action Report," 2014.

[30] J. G. J. (PBL) Olivier, G. (EC–J. Janssens–Maenhout, M. (EC–J. Muntean, J. A. H. W. (PBL) Peters, "Trends in Global CO_2 Emissions: 2016 Report," *PBL Netherlands Environ. Assess. Agency Eur. Comm. Jt. Res. Cent.*, pp. 86, 2016.

[31] Hacker F, Harthan R, Matthes F, Zimmer W. Environmental impacts and impact on the electricity market of a large scale introduction of electric cars in Europe–Critical Review of Literature," *ETC/ACC Tech. Pap.*, vol. 4, pp. 56-90, 2009.

[32] Conti S, Nicolosi R, Rizzo SA, Zeineldin HH. Optimal dispatching of distributed generators and storage systems for MV islanded microgrids. IEEE Transaction Power Delivery. 2012;**27**: 1243-1251.

[33] Etxeberria A, Vechiu I, Camblong H, Vinassa JM. Comparison of three topologies and controls of a hybrid energy storage system for microgrids. Energy Convers. Management. 2012;**54**:113-121.

[34] Department of Energy Office of Electricity Delivery and Energy Reliability. *Summary Report: 2012 DOE Microgrid Workshop*. 2014.

[35] Colet A, Ruiz A, Gomis O, Alvarez A, Sudria A. Centralized and distributed active and reactive power control of a utility connected microgrid using IEC61850. IEEE Syst. J.2012;**6**: 58-67.

[36] Aghamohammadi MR, Abdolahinia H. A new approach for optimal sizing of battery energy storage system for primary frequency control of islanded Microgrid. Int. J. Electr. Power Energy Syst.2014;**54**:325-333.

[37] Daneshi H, Khorashadi H. Microgrid energy management system: A study of reliability and economic issues. IEEE Power and Energy Society General Meeting, pp. 1-5, 2012.

[38] Trujillo CL, Velasco D, Figueres E, Garcerá G. Analysis of active islanding detection methods for grid–connected microinverters for renewable energy processing. Appl. Energy, 2010;**87**: 3591-3605.

[39] Ma T, Yang H, Lu L. A feasibility study of a stand–alone hybrid solar–wind–battery system for a remote island. Appl. Energy. 2014;**121**:149-158.

[40] Rajesh KS, Dash SS, Rajagopal R, Sridhar R. A review on control of AC microgrid, Renew. Sustain. Energy Rev. 2017;**71**:814-819.

[41] Zuo S, Davoudi A, Song Y, Lewis FL. Lewis. Distributed finite-time voltage and frequency restoration in islanded AC microgrids, IEEE Trans. Ind. Electron., 2016;**63**:5988-5997.

[42] Veneri O. *Technologies and Applications for Smart Charging of Electric and Plug-in Hybrid Vehicles*. 1st ed. Cham, Switzerland: Springer, 2017, pp. 39-64.

[43] Mohamad AMEI, Mohamed YARI, Investigation and assessment of stabilization solutions for DC microgrid with dynamic loads. IEEE Trans Smart Grid, 2019;**10**:5735-5747.

[44] Lotfi H, Khodaei A. AC versus DC microgrid planning, IEEE Trans. Smart Grid, 2017;**8**:296-304.

[45] Ma T, Cintuglu MH, Mohammed O.A. Control of a hybrid AC/DC microgrid involving energy storage and pulsed loads, IEEE Trans. Ind. Appl., 2017 ;**53**:567-575.

[46] Mehrizi-Sani A. Iravani R. Potential-function based control of a microgrid in islanded and grid-connected modes. IEEE Trans. Power Syst., 2010; **25**:1883-1891.

[47] Alsaidan I, Khodaei A, Gao W. A comprehensive battery energy storage optimal sizing model for microgrid applications. IEEE Trans. Power System. 2018;**33**:3968-3980.

[48] Ibrahim H, Belmokhtar H, Ghandour M. Investigation of usage of compressed air energy storage for power generation system improving—Application in a microgrid integrating wind energy. Energy Procedia, 2015;**73**:305-316.

[49] Arani AAK, Karami H, Gharehpetian HB, Hejazi MSA. Review of flywheel energy storage systems structures and applications in power systems and microgrids. Renewable Sustainable Energy Reviews. 2017; **69**:9-18.

[50] Inthamoussou FA, Pegueroles J, Bianchi FD. Control of a supercapacitor energy storage system for microgrid applications. IEEE Trans. Energy Convers. 2013;**28**:690-697.

[51] Nguyen TT, Yoo HJ, Kim HM. Applying model predictive control to SMES system in microgrids for eddy current losses reduction. IEEE Trans Appl. Superconduct. 2016; **26**:1-25.

[52] Konstantinopoulos SA, Anastasiad AG, Vokas GA, Kondylis GP, Polyzakis A. Optimal management

of hydrogen storage in stochastic smart microgrid operation. Int. J. Hydrogen Energy. 2017;**43**:490-499.

[53] Hannan MA, Hoque MM, Mohamed A, Ayob A. Review of energy storage systems for electric vehicle applications: Issues and challenges. Renew. Sustain. Energy Review. 2017;**69**:771 –789.

[54] Oriti G, Julian AL, Anglani N, Herna ndez GD. Novel hybrid energy storage control for single-phase energy management system in a remote islanded microgrid. in *Proc. IEEE Energy Conversion Congr. Expo. (ECCE)*, Oct. 2017, pp. 1552-1559.

[55] Guney MS, Tepe Y. Classification and assessment of energy storage systems. Renew Sustain. Energy Reviews. 2017;**75**:1187-1197.

[56] Garimela N, Nair NKC. Assessment of battery energy storage systems for small–scale renewable energy integration. In: IEEE TENCON, pp. 1-6, 2009.

[57] Chen H, Cong TN, Yang W, Tan C, Li Y, Ding Y. Progress in electrical energy storage system: A critical review. Prog. Nat. Sci., 2009;**19**:291-312.

[58] Hadjipaschalis I, Poullikkas A, Efthimiou V. Overview of current and future energy storage technologies for electric power applications. Renew. Sustainable Energy Reviews., 2009;-**13**:1513-1522.

[59] Abdin Z, Khalilpour KR. Single and Polystorage Technologies for Renewable based Hybrid Energy Systems. In Polygeneration with Polystorage for Chemical and Energy Hubs, pp. 77-131, 2019.

[60] Faisal M, Hannan MA, Ker PJ, Hussain A, Mansor MB, Blaabjerg F. Review of energy storage system technologies in microgrid applications: Issues and challenges. IEEE Access, 2018; **6**:35143-35164.

[61] Luo X, Wang J, Dooner M, Clarke J. Overview of current development in electrical energy storage technologies and the application potential in power system operation. Appliyng Energy, 2015;**137**:511-536.

[62] Liu H, He Q, Borgia A, Pan L, Oldenburg CM. Thermodynamic analysis of a compressed carbon dioxide energy storage system using two saline aquifers at different depths as storage reservoirs. Energy Convers. Manage., 2016;**127**:149-159.

[63] Yao E, Wang H, Wang L, Xi G, Maréchal F. Thermoeconomic optimization of a combined cooling, heating and power system based on small scale compressed air energy storage. Energy Convers. Manage., 2016;**118**: 377-386.

[64] Niaz S, Manzoor T, Pandith AH. Hydrogen storage: Materials, methods and perspectives. Renew. Sustain. Ene Reviews., 2015; **50**:457-469.

[65] Bhandari R, Trudewind CA, Zapp P. Life cycle assessment of hydrogen production via electrolysis—A review. Journal Cleaner Prod., 2014; **85**:151-163.

[66] Kousksou T, Bruel P, Jamil A, El Rhafiki T, Zeraouli Y. Energy storage: Applications and challenges. Sol. Energy Mater. Sol. Cells, 2014;**120**: 59-80.

[67] Kusko A, DeDad J. Stored energy–short–term and long–term energy storage methods for standby electric power systems. IEEE Trans. Industrial Applications. 2007;**13**:66-72.

[68] Crider JM, Sudhoff SD. Reducing impact of pulsed power loads on microgrid power systems. IEEE Trans. Smart Grid. 2010;**1**: 270-277.

[69] Farhadi M, Mohammed O. Energy storage technologies for high-power applications. IEEE Trans. Ind. Appl, 2016;**52**:1953-1961.

[70] Zhang X, Zhang Z, Pan H, Salman W, Yuan Y, Liu Y. A portable high-efficiency electromagnetic energy harvesting system using supercapacitors for renewable energy applications in railroads. Energy Convers. Manage., 2016;**118**:287-294.

[71] Dubal DP, Ayyad O, Ruiz V, Gómez P. Hybrid energy storage: The merging of battery and supercapacitor chemistries. Chem. Soc. Rev., 2015; **44**:1777-1790.

[72] Liu Q, Nayfeh MH, Yau ST. Supercapacitor electrodes based on polyaniline–silicon particle composite. Journal Power Sources, 2010;**195**: 3956 –3959.

[73] Gong K, Shi J, Liu Y, Wang Z, Ren L, Zhang Y. Application of SMES in the microgrid based on fuzzy control. IEEE Trans. Appl. Supercond. 2016; **26**:1-5.

[74] Ridgers TJ, Boucey C, Frambach JP, et al., Challenges in integrating distributed energy storage systems into future smart grid. IEEE radio and Wireless, Symposium, 2008, pp. 547-550.

[75] Molina MG, Mercado PE. Power flow stabilization and control of microgrid with wind generation by superconducting magnetic energy storage. IEEE Trans. Power Electron. 2011; **26**:910-922.

[76] Chen H, Cong TN, Yang W, Tan C, Li Y, Ding Y. Progress in electrical energy storage system: A critical review. Prog. Nat. Sci., 2009;**19**:291-312.

[77] Walsh F. Progress & challenges in the development of flow battery Technology. In: The 1st Int. Flow Battery Forum (IFNF): pp. 1-8, 2010.

[78] Xu L, Miao Z, Fan L, Gurlaskie G. Unbalance and Harmonic Mitigation using Battery Inverts. IEEE 2015 North American Power Symposium (NAPS), Charlotte, USA, 2015.

[79] Faisal M, Hannan MA, Ker PJ, Hussain A, Mansor MB, Blaabjerg F. Review of energy storage system technologies in microgrid applications: Issues and challenges. IEEE Access, 2018;**6**:35143-35164.

[80] Arghandeh R, Pipattanasomporn M, Rahman S. Flywheel Energy Storage System for Ride–Through Applications in a Facility Microgrid. IEEE Trans. on Smart Grid., 2012;**3**:1955-1962.

[81] Samineni S, Johnson BK, Hess HL, Law JD. Modeling and analysis of a flywheel energy storage system for voltage sag correction. IEEE Trans. Ind. Appl., 2006;**42**:42-52.

[82] Westering W, Hellendoorn H. Low voltage power grid congestion reduction using a community battery: design principles, control and experimental validation. Electrical Power and Energy Systems. 2020;**114**:1-9.

[83] Igrahim H, Belmokhtar K, Ghandour M. Investigation of usage of Compressed Air Energy Storage for Power Generation System Improving–Application in a Microgrid Integrating Wind Energy. 9th Int. Renewable Energy Storage Conf., IRES, pp. 305-316, 2015.

[84] Goharshenasan P, Joorabian M, Seifossadat SG. A new proposal for the design of hybrid AC/DC microgrids toward high power quality. Turkish Journal of Electrical Engineering & Computer Sciences, 2017; **25**:4033-4049.

[85] Hemmati R, Saboori H. Emergence of hybrid energy storage systems in renewable energy and transport applications—A review. Renew. Sustain. Energy Reviews., 2016; **65**:11-23.

[86] Dou X, Quan X, Wu Z, Hu M, Sun J, Yang K, Xu M. Improved Control Strategy for Microgrid Ultracapacitor Energy Storage System. Energy, 2014;**7**:8095-8115.

[87] Sharma R, Suhag S. Supercapacitor utilization for power smoothening and stability improvement of a hybrid energy system in a weak grid environment. Turkish Journal of Electrical Engineering & Computer Sciences, 2018;**26**:347-362.

[88] Cericola D, Novák P, Wokaun A, Køtz R. Hybridization of electrochemical capacitors and rechargeable batteries–an experimenttal analysis. *J. Power* Sources, 2011;**196**:10305-10313.

[89] Peng, et. al., Application of superconducting magnetic energy storage in microgrid containing new energy. IOP Conference series: Materials Science and Engineering, vol. 382, 2018.

[90] Morandi A, Trevisani L, Negrini F, Ribani PL, Fabbri M. Feasibility of superconducting magnetic energy storage on board of ground vehicles with present state of the art superconductors. IEEE Transaction Appl. Supercond., 2012;**22**:1558-1569.

[91] Suzuki S, Baba J, Shutoh K, Masada E. Effective applications of superconducting magnetic energy storage (SMES) to load leveling for high speed transportation system. IEEE Trans. Appl. Supercond., 2004;**14**: 713-716.

[92] Teleke S, Baran M, Huang A, Bhattacharya S, Anderson L. Control strategies for battery energy storage for wind farm dispatching. IEEE Trans. on Energy Convers., 2009;**24**: 725 –732.

[93] Nelson R. Power requirements for batteries in hybrid electric vehicles. J. Power Sources, 2000;**91**:2-26.

[94] Latha R, Palanivel S, Kanakaraj J. Frequency control of Microgrid based on Compressed Air Energy Storage System. Distribute Generation and Alternative Energy Journal, 2013;**27**:8-19.

[95] Shi R, Zhang X, Xu H, Liu F, Li W, Mao F, Yu Y. A Method for Microgrid connected Fuel Cell Inverters Seamless Transitions between Voltage Source and Current Source Modes. 9th International Conference on Power Electronics–ECCE, Seoul, Korea, vol. 63, pp. 1-5, 2015.

[96] Vigneysh T, Kumarappan N, Arulraj R. Operation and control of wind/fuel cell based hybrid microgrid in grid connected mode. IEEE International Multi–Conference on Automation, Computing, Communication, Control and Compressed Sensing, India, pp. 754-758, 2013.

[97] Ngamroo I, Karaipoom T. Improving Low–Voltage Ride–Through Performance and Alleviating Power Fluctuation of DFIG Wind Turbine in DC Microgrid by Optimal SMES with Fault Current Limiting Function. IEEE Trans. on Appl. Superconduct., 2014; **24**:1-16.

[98] Bolborici V, Dawson FP, Lian KK. Hybrid energy storage systems: connecting batteries in parallel with ultracapacitors for higher power density. IEEE Ind. Applying Mag., 2014;**20**:31-40.

[99] Gao L, Douglas R, Liu S. Power enhancement of an actively controlled battery/ultracapacitor hybrid. IEEE Transaction Power Electronic, 2005;**20**:263-243.

[100] Ise T, Kita M, Taguchi A. A hybrid energy storage with a SMES and secondary battery. IEEE Trans. Appl. Supercond, 2005;**15**:1915-1918.

[101] Briat O, Vinassa JM, Lajnef W, Azzopardi S, Woirgard E. Principle, design and experimental validation of a

flywheel-battery hybrid source for heavy-duty electric vehicles. IET Electr Power Appl, 2007;**1**:665-674.

[102] Prodromidis GN, Coutelieris FA, Simulations of economic and technical feasibility of battery and flywheel hybrid energy storage systems in autonomous projects. Renewable Energy, 2012;**39**:149-153.

[103] Lemofouet S, Rufer A. A hybrid energy storage system based on compressed air and supercapacitors with maximum efficiency point tracking. IEEE Trans. Ind. Electron., 2006;**53**:1105-1115.

[104] Zhao P, Dai Y, Wang J. Design and thermodynamic analysis of a hybrid energy storage system based on A – CAES (adiabatic compressed air energy storage) and FESS (flywheel energy storage system) for wind power application. Energy, 2014;**70**: 674-684.

[105] Wu Y, Gao H. Optimization of fuel cell and supercapacitor for fuel-cell Electric vehicles. IEEE Trans. Veh. Technol., 2006;**55**:1748-1755.

[106] Zhu GR, Loo KH, Lai YM, Tse CK. Quasi-maximum efficiency point tracking for direct methanol fuel cell in DMFC/supercapacitor hybrid energy system. IEEE Trans. Energy Convers., 2012;**27**:561-571.

[107] Martín IS, Ursúa A, Sanchis P. Integration of fuel cells and supercapacitors in electrical microgrids: Analysis, modelling and experimental validation. Int. Journal Hydrogen Energy, 2013; **38**:11655-11671.

[108] Loui H, Strunz K. Superconducting Magnetic Energy Storage for energy cache control in modular distributed hydrogenelectric energy systems. IEEE Trans. Appl. Supercond., 2007; **17**:361-364.

[109] Lee H, Shin BY, Han S, Jung S, Park B, Jang G. Compensation for the power fluctuation of the large scale wind farm using hybrid energy storage applications. IEEE Trans. Appl. Supercond., 2012;**22**: 57019.

[110] Zandi M, Payman A, Martin JP, Pierfederici S, Davat B, MeibodyTabar F. Energy management of a fuel cell/ supercapacitor/battery power source for electric vehicular applications. IEEE Trans. Veh. Technol., 2011;**60**: 433-443.

[111] Palizban O, Kauhaniemi K, Guerrero J. M. Microgrids in active network management—Part II: System operation, power quality and protection, Renew. Sustain. Energy Rev. 2014;**36**: 440-451.

[112] Li YW, He J. Distribution system harmonic compensation methods An overview of DG-interfacing inverters. IEEE Ind. Electron. Mag., 2014;**8**:18-31.

[113] Kow K W, Wong YW, Rajkumar RK. Power quality analysis for PV grid connected system using PSCAD/EMTDC. Int. J. Renew. Energy Res., 2015;**5**:121 –132.

[114] Hu WX, Xiao XY, Zheng ZX. Voltage sag/swell waveform analysis method based on multidimension characterization. IET Gener., Transmiss. Distrib., 2020;**14**:486-493.

[115] Zheng F, Chen Y, Zhang Y, Lin Y, Guo M. Low voltage ride through capability improvement of microgrid using a hybrid coordination control strategy. Journal Renew. Sustain. Energy, 2019;**11**:034102.

[116] Erlich I, Bachmann U. Grid code requirements concerning connection and operation of wind turbines in Germany. in *Proc. IEEE Power Eng. Soc. Gen. Meeting*, Jun. 2005:1253-1257.

[117] Alwaz N, Raza S, Ali S, Bhatti MKL, Zahra S. Harmonic power sharing and power quality improvement of droop controller based low voltage islanded microgrid, presented at the Int. Symp. Recent Adv. Electr. Eng. (RAEE), Aug. 2019, pp. 1-6.

[118] Astorga OAM, Silveira JL, and J. C. Damato, The influence of harmonics from non-linear loads in the measuring transformers of electrical substations, in *Laboratory of High Voltage and Electric Power Quality, Optimization Group of Energy Systems*, vol. 1, no. 4. São Paulo, Brazil: Sao Paulo State Univ., 2006, pp. 275-280.

[119] Blooming TM, Carnovale DJ. Application of IEEE STD 519-1992 harmonic limits, in *Proc. Conf. Rec. Annu. Pulp Paper Ind. Tech. Conf.*, Jun. 2006, pp. 1-9.

[120] Cho N, Lee H, Bhat R, Heo K. Analysis of harmonic hosting capacity of IEEE Std. 519 with IEC 61000-3-6 in distribution systems, presented at the IEEE PES GTD Grand Int. Conf. Expo. Asia (GTD Asia), 2019, pp. 730-734.

[121] Cleveland FM. IEC 61850-7-420 communications standard for distributed energy resources (DER), presented at the IEEE Power Energy Soc. Gen. Meeting-Convers. Del. Elect. Energ 21st Century, Jul. 2008, pp. 1-4.

[122] *IEEE Standard for Interconnecting Distributed Resources with Electric Power Systems-Amendment 1*, IEEE Standard 1547, Inst. Elect. Electron. Eng., New York, NY, USA, 2014.

[123] Wu YK, Lin JH, Lin HJ. Standards and guidelines for grid–connected photovoltaic generation systems: A review and comparison. IEEE Trans. Ind. Appl., 2017;**53**:3205-3216.

[124] Gao DW, Muljadi E, Tian T, Miller M, Wang W. Comparison of standards and technical requirements of grid-connected wind power plants in China and the United States. Nat. Renew. Energy Lab., Golden, CO, USA, Tech. Rep. NREL/TP-5D00-64225, 2016.

[125] Kim YJ. Development and analysis of a sensitivity matrix of a three–phase voltage unbalance factor, IEEE Trans. Power Syst., 2018;**33**:3192-3195.

[126] Savaghebi M, Jalilian A, Vasquez JC, Guerrero JM, Secondary control scheme for voltage unbalance compensation in an islanded droopcontrolled microgrid, IEEE Trans. Smart Grid, 2012;**3**:797-807.

[127] State Grid Corporation of China. Technical rule for PV power station connected to power grid, Chin. Enterprise Standards, Tech. Rep. GB/T 19964, 2012.

[128] Wu YK, Lin JH, Lin HJ. Standards and guidelines for grid–connected photovoltaic generation systems: A review and comparison, IEEE Trans. Ind. Appl., 2017;**53**:3205-3216.

[129] Troester E. New German grid codes for connecting PV systems to the medium voltage power grid, presented at the 2nd Int. Workshop Concentrating Photovoltaic Power Plants, Opt. Design, Prod., Grid Connection, 2009, pp. 9-10.

[130] *Interconnection of Distributed Resources and Electricity Supply Systems*, Canadian Standards Association, Standard CSA C22.3 No. 9-08-R2015, 2015. Accessed: Mar. 29, 2020. [Online]. Available: https://www.csagroup.org

[131] Ghassemi F, Perry M. *Review of Voltage Unbalance Limit in the GB Grid Code CC.6.1.5 (b)*. Accessed: March. 11, 2021. [Online]. Available: https://www.nationalgrid.com.

[132] Xu L, Miao Z, Fan L, Gurlaskie G. Unbalance and Harmonic Mitigation

using Battery Inverts. IEEE 2015 North American Power Symposium (NAPS), Charlotte, USA, 2015.

[133] Li D, Zhu ZQ. A Novel Integrated Power Quality Controller for Microgrid. IEEE Trans. on Ind. Electron., 2015;**62**:2848-2858.

[134] Bajpai RS, Gupta R. Voltage and power flow control of grid connected wind generation system using D-STATCOM. IEEE Power Energ Soc. Gen. Meeting—Convers. 21st *Century*, Jul. 2008, pp. 1-6.

[135] Castilla M, Miret J, Matas J, de Vicuna LG, Guerrero JM. Linear current control scheme with series resonant harmonic compensator for single phase grid connected photovoltaic inverters. IEEE Transactions on Industrial Electronics, 2008;**55:** 2724-2733.

[136] Ghahderijani MM, Castilla M, de Vicuña LG, Camacho A, Martínez JT. Voltage sag mitigation in a PV-based industrial microgrid during grid faults, presented at the 26th IEEE Int. Symp. Ind. Electron. (ISIE), 2017, pp. 186-191.

[137] Li Z, Li W, Pan T. An optimized compensation strategy of DVR for microgrid voltage sag. Protection Control Mod. Power Syst, 2016;**1**:1-8.

[138] Al-Shetwi AQ, Sujod MZ. Modeling and control of grid–connected photovoltaic power plant with fault ride through capability. J. Sol. Energy Eng., 2018;**140**:021001-021009.

[139] Molinas M, AreSuul J, Undeland T. Low voltage ride through of wind farm with cage generators: STATCOM versus SVC. IEEE Trans. Power Electron., 2008;**23**:1104-1117.

[140] Lee TL, Hu SH, Chan YH. DSTAT-COM with positive-sequence admittance and negative-sequence conductance to mitigate voltage fluctuations in high-level penetration of distributed-generation systems, IEEE Trans. Ind. Electron., 2013;**60**: 1417-1428.

[141] Chaudhari P, Rane P, Bawankar A, Shete P, Kalange K, Moghe A, Panda J, Kadrolkar A, Gaikwad K, Bhor N, Nikam V. Design and implementation of STATCOM for reactive power compensation and voltage fluctuation mitigation in microgrid, presented at the IEEE Int. Conf. Signal Process., Informat., Commun. Energy Syst. (SPICES), Feb. 2015, pp. 1-5.

[142] Goyal M., John B, Ghosh A. Harmonic mitigation in an islanded microgrid using a DSTATCOM, presented at the IEEE PES Asia–Pacific Power Energy Eng. Conf. (APPEEC), Nov. 2015, pp. 1-5.

[143] Yang X, Du Y, Su J, Chang L, Shi Y, Lai J. An optimal secondary voltage control strategy for an islanded multibus microgrid, IEEE J. Emerg. Sel. Topics Power Electron., 2016;**4**: 1236-1246.

[144] Bajpai RS, Gupta R. Voltage and power flow control of grid connected wind generation system using DSTAT-COM. *IEEE* Power Energy Soc. Gen. Meeting—Convers. 21st *Century*, Jul. 2008, pp. 1-6.

[145] Castilla M, Miret J, Matas J, de Vicuna LG, Guerrero JM. Linear current control scheme with series resonant harmonic compensator for single phase grid connected photovoltaic inverters. IEEE Transactions on Industrial Electronics, 2008;**55:** 2724-2733.

[146] Khadem SK, Basu M, Conlon MF. Integration of UPQC for power quality improvement in distributed generation network—A review, presented at the 2nd IEEE PES Int. Conf. Exhib. Innov. Smart Grid Technol., Dec. 2011, pp. 1-5.

[147] Rasheed A, Keshava G. Improvement of power quality for

microgrid using fuzzy based UPQC controller, Indian J. Sci. Technol., 2015;**8**:1-5.

[148] Senthil Kumar A, Rajasekar S, Raj PADV. Power quality profile enhancement of utility connected microgrid system using ANFIS-UPQC, Procedia Technol., 2015;**21**:112-119.

[149] Singh MD, Mehta RK, Singh AK. Performance assessment of current source converter based UPQC for power quality improvement with simple control strategies, J. Elect. Syst., 2019;**15**:276-290.

[150] Hoseinnia S, Akhbari M, Hamzeh M, Guerrero JM. A control scheme for voltage unbalance compensation in an islanded microgrid, Electric Power Syst. Res., 2019;**177**: 106016.

[151] Dai L, Chen W, Yang X, Zheng M, Yang Y, Wang R. A Multi-Function Common Mode Choke Based on Active CM EMI Filters for AC/DC Power Converter. IEEE Access, 2019;**7**: 43534-43546.

[152] Hernandez E, Madrigal M. A step forward in the modeling of the Doubly Fed Induction Machine for harmonic analysis. IEEE Trans. on Energy Conv., 2014;**29**:149-157.

[153] Baradarani F, Zadash E, Zamani MA. A Phase-Angle Estimation Method for Synchronization of Grid Connected Power Electronic Converters. IEEE Trans. on Power Deliv., 2015;**30**:827-835.

[154] Senthilkumar A, et. al., Mitigation of Harmonic Distortion in Microgrid System Using Neural Learning Algorithm Based Shunt Active Power Filter. Smart Grid Technol. Elsevier, 2015;**21**:147-154.

[155] Mehta G, Singh S. Power quality improvement through grid integration of renewable energy sources. IETE Journal of Research, 2013;**59**: 210-218.

[156] Ling Q, Lu Y. An Integration of supercapacitor storage research for improving low voltage ride through in power grid with wind turbine. In: Presented at the power and energy engineering conference, APPEEC Shanghai, China: Asia-Pacific; 2012.

[157] Wang MQ, Gooi HB. Spinning reserve estimation in microgrids. IEEE Trans. Power Systems, 2011;**26**: 1164-1174.

Section 4

A Study on Micro-Grids toward 100% Renewable Energy

Chapter 4

An Overview Study of Micro-Grids for Self-Production in Renewable Energies

Hocine Sekhane

Abstract

Micro-grids (μ-grids) are small-scale power grids, specially designed to provide low voltage (LV) power supply to a small number of consumers. These networks include: different production units (energy resources), storage devices and local controllable loads, which have the possibility of being controlled. In this chapter, we will study in detail the constitution of an electrical micro-grid, their two operating modes (connected mode and islanded mode), and their controls. On the other hand, we will also discuss on hybrid micro-grids and their advantages. We will also discuss for the monitoring and data logging products used in micro-grids and hybrid micro-grids. Finally, at the end of this chapter we will ended with the importance of micro-grids systems.

Keywords: Self-production, Renewable energy, Micro-grid, Monitoring, islanded mode

1. Introduction

Since the advent of electricity and the establishment of its generating stations, its distribution has been mainly focused on urban and populated areas, as many rural and desert communities are still isolated from the larger traditional networks due to geographic and economic constraints. Providing electricity to rural and desert populations outside the global grid remains a major task for many developing and developed countries alike, and according to the International Energy Agency, micro-grids represent the most cost-effective solution to providing universal electricity access to these small (or micro) communities [1].

A micro-grid or μ-grid is defined as a group of distributed resource (DR) units which is designed to provide low voltage (LV) power supply to a small number of loads, and can operate in grid-connected mode, islanded (autonomous) mode (in the event of a fault in the main network), or ride-through between the two modes [2, 3]. This network includes [4, 5]:

1. Different local production units (energy resources) (micro-turbines, fuel cells, small diesel generators, photovoltaic panels, mini-wind turbines, small hydro).

2. Storage devices (flywheels energy storage (FES), energy capacitors and batteries).

3. Local controllable loads, which have possibilities of being controlled vis-a-vis the operation of the network.

In many areas, micro-grid provides an attractive alternative with improved stability when compared to centralized systems which are not feasible due to the relatively small loads scattered in remote areas. In addition, transmission is a problem for geographically isolated areas, and this makes off-grid alternatives very necessary in some situations. The large distances between rural and remote desert areas on the one hand and central generation centers on the other hand, make it possible to lose approximately 30% of the transmitted energy, which greatly reduces the efficiency of the overall electrical system. Therefore, local micro-grids with on-site generation provide a very reasonable alternative [6].

In what follows we will first explain the micro-grids operating system and there control, then we will present the hybrid micro-grids with distributed generation and accumulation, after we will discuss to the monitoring and data logging products such as: Consospy and Webdynsun technologies, and finally we highlight the importance of micro-grids systems.

2. Micro-grids operating and control system

Micro-grids can be connected directly to the LV distribution network or operate in islanded mode. In the field of renewable energies, an islanded system is an autonomous electricity production system operated to supply consumers in isolated regions (without access to the public electricity grid). The Chernobyl accident of April 25, 1986 occurred during an islanding test [7].

In order to achieve long-term island operation, a micro-grid must meet high requirements in terms of storage size and nominal capacity of micro-generators for a continuous supply of all loads on which it must rely great flexibility on demand.

Generally, the maximum capacity of a μ-grid in terms of peak load demand is limited to a few MW, but other regions may have different upper limits [7].

A μ-grid has 3 essential characteristics: local load, local μ-sources and intelligent control. The following are misconceptions regarding μ-networks:

- μ-grids are exclusively isolated systems.
- Customers who own μ-sources build a μ-grid.
- μ-grids are composed of intermittent renewable energy sources (RES), so they must be unreliable and easily prone to blackouts.
  Clarification: A μ-grid can compensate for fluctuation in renewable energy sources through its own storage units (when is in islanded mode) or external production reserves (when is connected to the grid). In addition, the ability of the μ-grid to switch from connected mode to islanded mode actually improves security of supply.
- As μ-grids are expensive to build, the concept will be limited to field tests or only to remote sites.

The final schemas of configuration and exploitation of a micro-grid depend on potentially conflicting interests between the different stakeholders involved in the supply of electricity, such as system/grid operators, distributed generation (DG) owners, distributed generation operators, energy suppliers, customers and regulatory agencies. Therefore, optimal planning of operations in micro-grids can have economic, technical and environmental objectives [7].

In the economic option, the objective is to minimize the total costs regardless of the impact/performance of the network. This option can be considered by owners or decentralized generation operators. Decentralized productions are exploited without worrying about network or emission obligations. The main limitations come from the physical constraints of distributed generation (DG).

The technical option optimizes the operation of the network (minimization of power losses, voltage variations and device load), without taking into account the costs and production revenues of distributed generation. This option may be preferred by system operators.

The environmental option performs DG units with lower emission levels, without taking into account economic or technical aspects. This is preferable to achieve environmental goals.

The combined objective option solves an optimal distribution problem of multi objective DG, taking into account all economic, technical and environmental factors.

The control of intermittent RES units (e.g. use of solar energy source in sunny weather and wind source in time away from the sun) is limited by the physical nature of the primary energy source. It is generally not advisable to reduce intermittent SER units unless they are causing line overloads or overvoltage issues.

The main control functionalities in a micro-grid can be distinguished into three groups [5]:

2.1 Upstream network interface

The main interaction with the upstream grid is linked to market participation, more specifically to micro-grid actions to import or export energy following decisions of the energy service company. Due to the relatively small size of a micro-grid, the energy service company can manage a larger number of micro-grids, in order to maximize its profits and provide ancillary services to the upstream grid [5].

2.2 Micro-grid internal control

This level includes all the functions of the micro-grid which require the collaboration of more than two actors. The functions at this level are as follows [5]:

- Forecast of load and renewable energy sources,
- Load shedding/management,
- Secondary voltage/frequency control,
- Secondary control of active/reactive power,
- Security monitoring.

2.3 Local control

This level includes all the local functions [5]:

- Protection functions,
- Primary voltage/frequency control,

- Active/reactive primary power control,
- Battery management.

On the other hand, and due to the intermittent nature of Renewable Energy Sources (RES) that greatly affect the operation of micro-grids systems, as well as the continuous fluctuations in demand of the loads, and in order to ensure and support its reliability, the achievement of equilibrium lies in the use of hybrid micro-grids that combine two or more technologies for the production of decentralized electrical energy. This is what makes it one of the best options available due to its many technical and economic advantages [8].

3. Hybrid micro-grids with distributed generation and accumulation

Today, modern technology allows the use of hybrid μ-grids which provide the generation and distributed storage (accumulation) of electricity. These hybrid μ-grids combine at least two technologies for power generation, and typically use renewable energy as primary energy source and diesel fuel as an auxiliary. This results in reliable, sustainable and profitable energy [9]. A hybrid micro-grid structure is depicted in **Figure 1** below.

As shown in figure above, we can see that this hybrid micro-grid combines two renewable energy technologies for power generation (solar and wind generation) as well as a number of diesel generators.

As an example, the "Princess Elisabeth" polar station depicted in **Figure 2** below is a scientific research station not connected to an electricity network because it is located in Antarctica in extreme climatic conditions (Air temperatures: –5° C to –50° C, maximum wind speed per month: 125 km/h) [10].

Thanks to the installation of a hybrid micro-grid, the station is energy self-sufficient. To produce electricity for the polar station, this hybrid μ-grid combines: solar

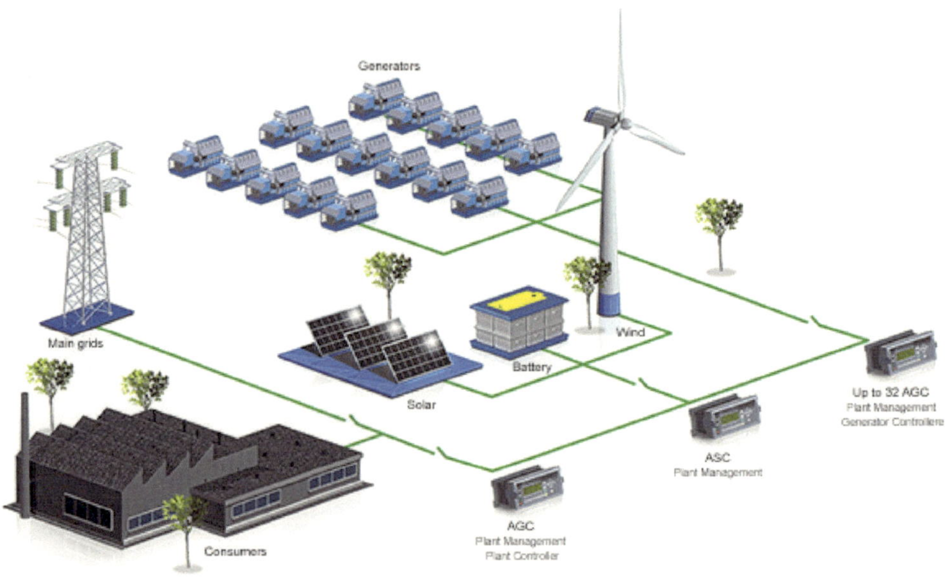

Figure 1.
Hybrid μ-grids.

Figure 2.
"Princess Elisabeth" polar station powered by a hybrid μ-grid.

panels (379.5 m^2) and wind turbines (9 wind turbines of 6 kWh each), then stored in lead-acid batteries with a capacity of 6 000 Ah. The heating is produced by thermal solar panels (22 m^2). Two diesel generators (44 kWh) are available as backup [10].

On the other hand, for proper functioning and better operation of micro-grids and hybrid micro-grids, it is highly necessary to integrate monitoring and data logging installations.

4. Monitoring and data logging

Network monitoring provides the information that network administrators need to determine whether the network is operating optimally in real time. Using tools like network monitoring software, administrators can proactively identify shortcomings, improve efficiency, and more [11].

There are different solutions to monitor the production and proper functioning of inverters, photovoltaic panels, etc. What we call them "monitoring", "data logger", and their function is to help us acquire and analyze the production data of solar panels, inverters, etc.

The role of monitoring and data logger devices can be summarized in the following points [9]:

- Real-time display of parameters
- Control and data logging.
- Creation of databases
- Logging and consultation of historical data, stored on computer, in the form of graphs or tables.
- Export to text files and spreadsheets

- Access to information through a simple internet browser
- Management and control of events or facts programmed by the user.
- Design of reports or simulation of electric bills for the allocation of partial costs.

Among the examples of monitoring and data logging products:

4.1 Consospy

The CONSOSPY Electricity module (see **Figure 3**) is a box that connects to the electronic electricity meter. Thanks to its storage capacity, it records the power of the counter at regular intervals (every minute, 10 minutes or every hour). With this module we can, thus, follow either our consumption, or our production of electricity. Communication is wireless (radio waves). The energy evolution of the different periods can be consulted with the monitoring software "SuiviConsoSpy" [12].

The operating principle of this module is schematized in **Figure 4**. All communication with the Internet module is carried out by radio. So there are no wires to install or holes to achieve. Operating on mains but also by batteries, the module is immune to power cuts! [12].

The energy evolution of the different periods can be viewed on the ConsoSpy Monitoring website from a smartphone, a tablet or computer without geographic limitation [12].

4.2 Webdynsun

The WebdynSun gateway makes it possible to monitor and collect data from a photovoltaic installation. On a single box, the gateway pools all the indicators coming from inverters, electricity meters and environmental sensors (sunshine, temperature, etc.).

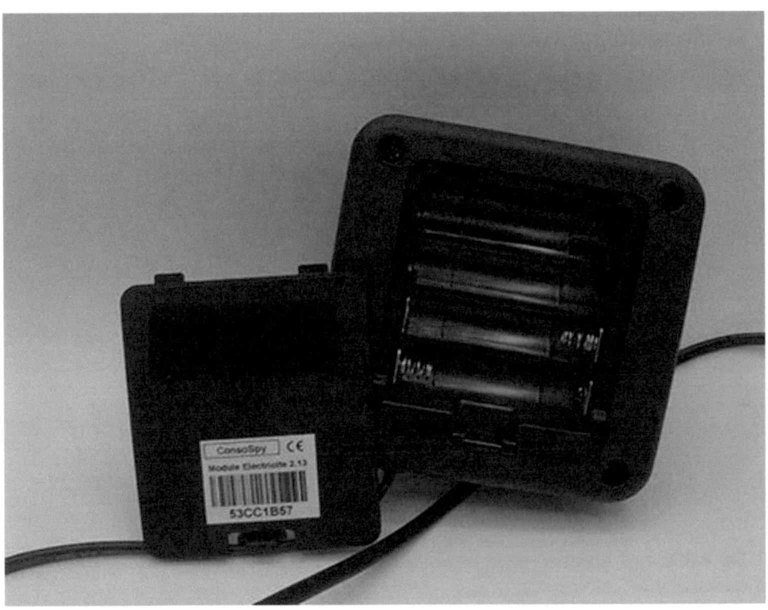

Figure 3.
Consospy electricity module.

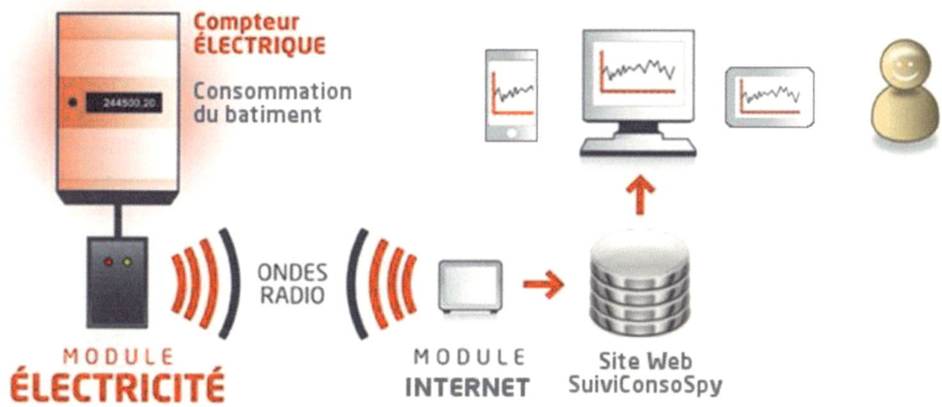

Figure 4.
Schema of the operating principle of the Consospy electricity module.

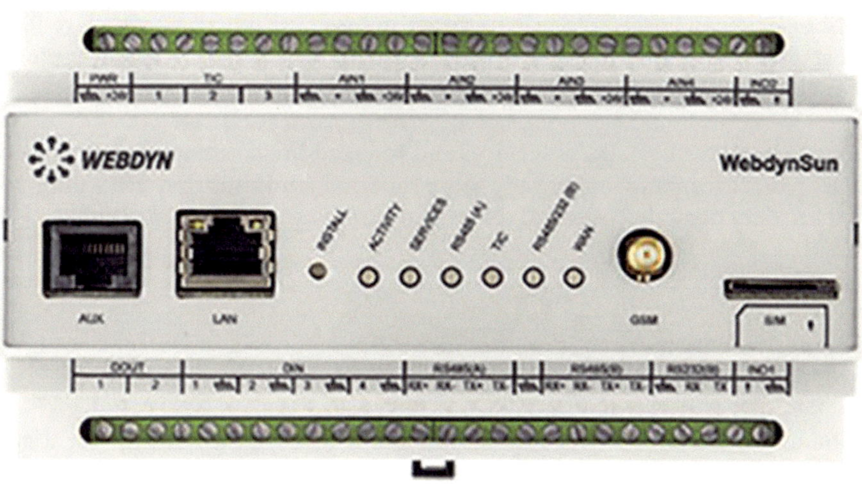

Figure 5.
Photo of a WebdynSun module (gateway).

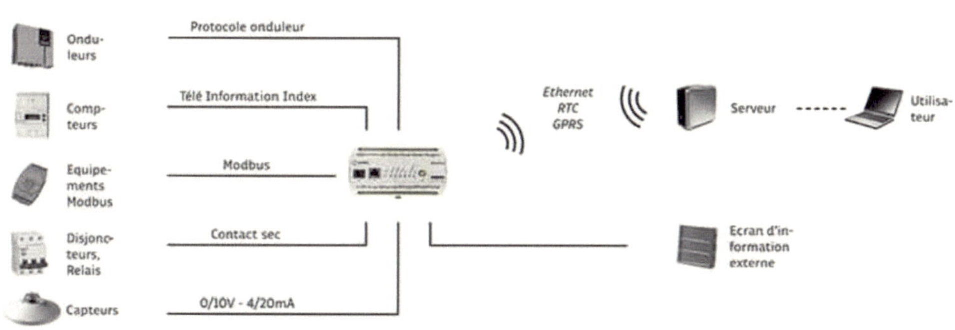

Figure 6.
Schema of the operating principle of the WebdynSun electricity module.

The objectives are preventive and curative remote maintenance of the plant as well as real-time monitoring of electricity production [13].

Figure 5 below represents a photo of the Webdynsun electricity module.

On the other hand, the schema of the operating principle of this module is shown in **Figure 6**. The WebdynSun gateway operates in an advanced data logging mode. From a configuration file and (or) from the local HTML interface, which describes all of the plant's equipment (inverters, meters, sensors, etc.), the WebdynSun gateway scans and collects the data associated with each equipment. These data are formatted and sent periodically, through the GPRS, Ethernet or telephone network to a federating server [13].

5. Importance of micro-grids systems

Micro-grids advantages can be summarized into technical, environmental, social and financial benefits as follows [14]:

5.1 Financial advantages

- Low fuel cost

- Micro-grid can spread electrical storage across multiple users which reduces cost when compared to home off-grid systems where electrical storage is concentrated in one area.

- Due to improved electrical services and reduced breakdowns such as power outages, customers are generally more satisfied, and thus they are willing to pay for services provided by small networks, resulting in increased revenues [14].

5.2 Technical advantages

- Micro-grids are more efficient because they can provide a low load at night when less electricity is required.

- Unlike conventional power generation, micro-grids reduce energy lost at night when society needs less energy as larger electrical systems such as diesel generators cannot provide this because they are ineffective at lower loads and often continue to operate at higher loads regardless of the amount of electricity required.

- The use of micro-grids reduces the amount of time the generators are run at low loads and thus increases the efficiency of the entire system.

- Micro-grids require less maintenance than large electrical networks. Since it reduces the hours of use of diesel generators at lower loads, the generators last longer and do not need to be replaced as often [14].

5.3 Social advantages

- In order for many companies and institutions to operate, they must have working and efficient electricity, so the micro-grids provide the necessary services for these companies and institutions to achieve success in developing regions and this leads to the creation of more job opportunities and increased income for society.

- Electricity micro-grids provide more opportunities for social gatherings and events that enhance the community and also create the opportunity to build more buildings and expand the community [14].

5.4 Environmental advantages

- Micro-grids are much more environmentally friendly than other types of grids because they reduce the need for diesel generators.

- Micro-grids reduce greenhouse gas emissions dramatically, and this reduces air pollution.

- Micro-grids reduce noise in areas where they are used [14].

6. Conclusion

In this chapter, we provided an overview of strategy of self-production in renewable energies or also called in another term "Micro-grids". Micro-grids are considered as systems that include LV distribution systems with distributed energy sources, storage systems and controllable loads. Through this chapter we have explained many concepts and principles such as: micro-grids operating systems, control of micro-grids, hybrid micro-grids with distributed generation and accumulation. We have also discussed for monitoring and data logging products such as Consospy and Webdynsun electricity modules and we ended with the importance of micro-grids systems.

Many excellent research works on various aspects related to micro-grids is done in Europe, United States, Japan and Canada where several activities were carried out such as: analysis of communication constraints and control architecture, development and improvement of micro source controllers intended for frequency and voltage control by droop curves, study of new market concepts for the sale of energy and micro-grid system services, the development of a centralized controller, ... etc.

Thanks

The author would like to thank the author service manager Kristina Kardum Cvitan for her helps.

Author details

Hocine Sekhane
University August 20, 1955 Skikda, Skikda, Algeria

*Address all correspondence to: docsekhoc@gmail.com

References

[1] Michael Franz, Nico Peterschmidt, Michael Rohrer, Bozhil Kondev: Guide pratique de la politique des mini réseaux. 2014.

[2] Katiraei F, Iravani M R: Power Management Strategies for a Microgrid With Multiple Distributed Generation Units. IEEE TRANSACTIONS ON POWER SYSTEMS. NOVEMBER 2006;21:1821-1831. DOI: 10.1109/TPWRS.2006.879260

[3] Haizea Gaztanaga Arantzamendi. Etude de structures d'intégration des systèmes de génération décentralisée: application aux micro-réseaux [thesis]. National Polytechnic Institute of Grenoble - INPG; 2006.

[4] Azeddine Houari. Contribution à l'étude de micro-réseaux autonomes alimentés par des sources photo voltaïques [thesis]. University of Lorraine; 2012.

[5] Nikos Hatziargyriou, editor. Microgrids: Architectures and control. 1st ed. IEEE press: Wiley; 2014. 341 p. ISBN: 978-1-118-72068-4

[6] Luiz Antonio de Souza Ribeiro, Osvaldo Ronald Saavedra, Shigeaki Leite de Lima, and José Gomes de Matos. Isolated Micro-Grids With Renewable Hybrid Generation: The Case of Lençóis Island. IEEE Transactions on Sustainable Energy (Volume: 2, Issue: 1, Jan. 2011). DOI: 10.1109/TSTE.2010.2073723

[7] Îlotage. [Internet]. 2019. Available from: https://www.france-science.org

[8] F. Badrkhani Ajaei; J. Mohammadi; G. Stevens; E. Akhavan. Hybrid AC/DC Microgrid Configurations for a Net-Zero Energy Community. 2019 IEEE/IAS 55th Industrial and Commercial Power Systems Technical Conference (I&CPS), 10 June 2019, Calgary, AB, Canada, DOI: 10.1109/ICPS.2019.8733323

[9] Daniel Cadilla, Mireia Gil, Cristian Ros, Cristina Gil, Nicola Bugati. Handbook of Micro-réseaux photovoltaïques hybrides: guide de conception et calcul. 1st ed. azimut360; 2017. 124 p.

[10] Micro-grids. [Internet]. 2020. Available from: http://www.smartgrids-cre.fr

[11] Basel Alsayyed, Hoda H. ElSheikh, Abbas Fadoun. Review of power quality monitoring systems. April 2015, DOI:10.1109/IEOM.2015.7093825

[12] Sébastien Bastard, Concepteur (logiciel et matériel) de la solution ConsoSpy. Manuel d'utilisation du logiciel SuiviConsoSpy 2.0

[13] WebdynSun - Manuel d'exploitation - Version 2.3

[14] James Hazelton, Anna Bruce, Iain Macgill. A review of the potential benefits and risks of photovoltaic hybrid mini-grid systems. January 2013, renewable energy. 67: 222-229. Doi: 10.1016/j.renene.2013.11.026

Section 5

Salp Swarm Optimization

Chapter 5

Salp Swarm Optimization with Self-Adaptive Mechanism for Optimal Droop Control Design

Mohamed A. Ebrahim, Reham M. Abdel Fattah, Ebtisam M. Saied, Samir M. Abdel Maksoud and Hisham El Khashab

Abstract

The collaboration of the various distributed generation (DG) units is required to meet the increasing electricity demand. To run parallel-connected inverters for microgrid load sharing, several control strategies have been developed. Among these methods, the droop control method was widely accepted in the research community due to the lack of important communication links between parallel-connected inverters to control the DG units within a microgrid. To help to solve the power-sharing process, keep to frequency and voltage constrained limits in islanded mode microgrid system. The parameter values must therefore be chosen accurately by using the optimization technique. Optimization techniques are a hot topic of researchers; hence This paper discusses the microgrid droop controller during islanding using the salp swarm inspired algorithm (SSIA). To obtain a better fine microgrid output reaction during islanding, SSIA-based droop control is used to optimally determine the PI gain and the coefficients of the prolapse control. The results of the simulation show that the SSIA-based droop control can control the power quality of the microgrid by ensuring that the keep to frequency and voltage constrained limits and deviation and proper power-sharing occurs during the microgrid island mode during a load change.

Keywords: Droop control, microgrid, salp swarm inspired algorithm, distributed generation, power-sharing

1. Introduction

In the last decade, the electricity demand was increase and shortly, the electricity demand will be expected to rise significantly [1]. To meet this projected demand, there is a trend towards renewable energy sources to be used because they are environmentally sound and are considered economically better [1]. This transition in electricity generation from conventional to renewable energy sources (RES) [2]. This has culminated in the development of small-scale power generation systems named microgrids [2]. A microgrid that involves local loads and Distributed generation sources (DGs) [3]. DG systems are ideal for highly reliable electrical power supply [3]. Various types of energy resources are currently available, such as wind turbines (WTs), photovoltaic systems (PVS), fuel cells (FC) [4]. It is difficult to connect these renewable resources directly to a utility grid [5]. To solve this problem, the microgrid is used to make the interface between the utility grid and distributed renewable resources [3]. Microgrids must be worked in the grid-connected mode as well as island mode contingency [3]. The power produced from most renewable resources is direct current (DC) but the utility grid is alternating current (AC). The inverter must be used to convert DC to AC. Therefore, an inverter is the main microgrid element [3]. In a microgrid, there are working Inverter parallel. The Inverters parallels are guaranteed to high reliability. Because if an inverter fails, the remaining modules can still supply the necessary power to the load [3]. The inverters control is intended to deliver the active and reactive energy while preserving the variability in frequency and voltage within the allowable limits [6]. To control inverters used the droop control technique. The droop control technique provides power-sharing, voltage and frequency constrained limits [7]. Such droop controllers are tuned with identical parameters in the d-axis and q-axis by trial and error method [8]. Nonetheless, in obtaining optimum parameters or even the right outcomes, this method has a major limitation.

Before human existence on this planet, nature used evolution to constantly solve challenging problems. The researchers inspired a solution based on nature to solve the difficult problems and challenges facing them. In 1977, Holland introduced a revolutionary idea in the field of optimization when evolutionary ideas in nature were modeled in computers to solve optimization problems [9]. The emergence of a new form of heuristic algorithm is Genetic Algorithms (GA), which is the most common and famous [10]. Opening the door for researchers to research and study to find new ways to solve the problems and challenges facing them in different fields.

Heuristic algorithms treat the problem as a black square with a combination of inputs and outputs. Their inputs are the problem variables and the outputs are the goals or objectives. A heuristic search begins with the formation of a collection of randomized inputs as the solution to the problem. The search is followed by analyzing each solution, monitoring objective values, and modifying /mixing/ developing output-based solutions. These steps will be repeated until solve the problem [9].

Researchers have a propensity to use optimization algorithms to overcome many engineering problems and challenges, but there is a question for researchers "Is there an only technique of optimization that can solve all problems? ". Lately, several types of algorithms have emerged such as Harris hawks optimization (HHO) [11], Salp Swarm Inspired Algorithm (SSIA) [12–14], grasshopper optimization algorithm (GOA) [1, 15], Sine Cosine Algorithm (SCA), Whale Optimization Algorithm (WOA) [16–18], Moth-Flame Optimization Techniques [19], Gray Wolf Algorithm (GWO) [20–22], Ant Lion Optimizer (ALO) [23], Moth-Flame Optimization algorithm (MFO) [24, 25], particle swarm optimization (PSO) [26, 27] and Dragonfly Algorithm (DA) [28] were the outcome of GA's success. Any heuristic algorithm is flawed and the output is influenced by its limitations.

But there's a fundamental question of "why do researchers keep discovering or developing new algorithms?". There is a theory that explains the answer to this question, which is called the rule of no free lunch (NFL) [23]. Logically, this principle shows that no one can suggest an algorithm to solve all optimization issues. This does not mean that the effectiveness of a type of algorithm in solving a particular set of problems is that it can solve all the optimization problems. The NFL principle enables researchers to suggest new optimization algorithms or to enhance existing algorithms to solve problem subtypes in various fields [23].

The study proposes an SSIA-based controller to optimize the parameters of the PI controller and drop control coefficients under load change conditions to control the voltage, frequency and power-sharing of an islanded MG.

2. Salp swarm inspired algorithm (SSIA)

There are several swarm algorithms, many of them inspired by food search behavior, that have emerged so far. A new swarm intelligence technique known as SSIA inspired by the action of salp was proposed by Mirjalili et al. [29]. Salps are from the Salpidae family. Salp is described as a body shaped like a barrel with a transparent body. Salps are strongly jellyfish-like. The salp environment in which he lives is difficult to reach and also difficult to provide this environment in the laboratory, so it is difficult to preserve. Biological studies of salp are therefore in its early stages. The salp forms the salp chain, a swarm that lives in the deep ocean [30]. The chains of salp are made up of two groups: leaders and followers. The leader is at the front of the chain, but the follower is named the remainder of salp. In an n-dimensional search space where n is the number of variables of a given problem, the location of salps is. It can be observed from the nature of the salp's actions that the leader salp goes around the source of food and the followers follow the leader. The leader changes his position in every iteration and the followers adopt it when finding food. In **Figure 1** The form and nature of the salp swarm.

Figure 2 illustrates the leader and follower's movement around the food. In a two-dimensional matrix named x, the location of all salps is stored. It is also believed that in the search space there is a food source called F as the target of the swarm. To alter the location of the Leader, the following equation is proposed [29]:

$$x_j^1 = \begin{cases} F_j + c_1((ub_j - lb_j)c_2 + lb_j)c_3 \geq 0 \\ F_j - c_1((ub_j - lb_j)c_2 + lb_j)c_3 < 0 \end{cases} \tag{1}$$

where
x_j^1: first Salp (leader) position in the jth dimension,
F_j: food source position of the jth dimension,
ub_j: upper bound of jth dimension,
lb_j: lower bound of jth dimension,
l: current iteration,
L: maximum number of iterations,
c_1, c_2, and c_3: random numbers uniformly generated in the interval of [0,1].

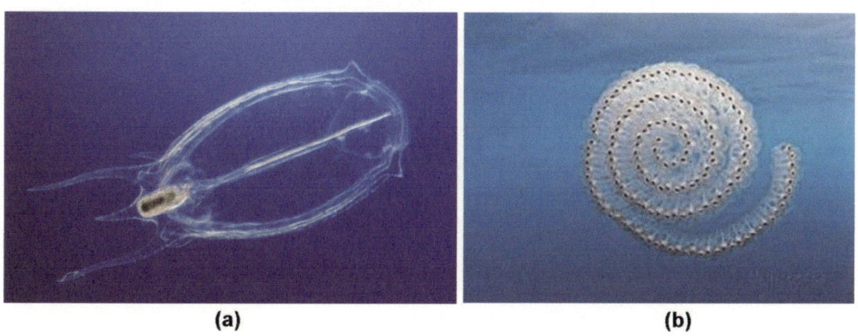

Figure 1.
Shape and structure of salp swarm in the deep ocean. (a) Single salp, and (b) single salp chain.

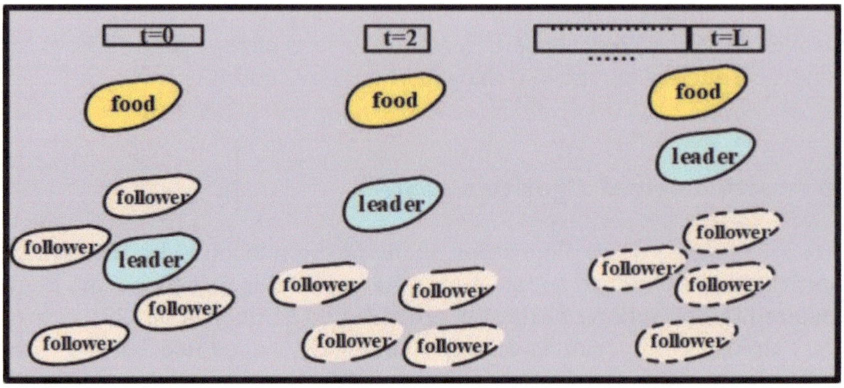

Figure 2.
The swarm of Salps (Salps chain) [31].

x^i_j: position of nth follower salp in ith dimension.

c_1 is a significant coefficient that balances exploration and exploitation. The following equation is used to estimate c_1 [12, 32]:

$$c_1 = 2e^{-\left(\frac{4l}{L}\right)^2} \tag{2}$$

Newton's law of motion will be used to update the position of the followers as follows:

$$x^i_j = \frac{1}{2}\, ct^2 + \lambda_0 t \tag{3}$$

Where I ≥ 2 and x^i_j shows the position of ith follower salp in jth dimension, t is time, λ_0 is the initial speed, and c = $\frac{\lambda_{final}}{\lambda_0}$ where λ = $\frac{x-x_0}{t}$.

For restructuring, in an optimization problem, it can be inferred that t is the iteration; this equation can be represented as follows:

$$x^i_j = \frac{1}{2}\left(x^i_j + x^{i-1}_j\right) \tag{4}$$

The advantages of SSIA:

1. A strong convergence acceleration.

2. Expedited method for providing excellent solutions.

3. Compatible with many types of optimization problems

4. A globally effective scheme to look for

5. Suitable for a broad search field.

6. In concept and implementation for related applications, SSA is simple.

7. A few parameters for tuning.

The SSIA pseudo-code algorithm is shown in **Figure 3**. **Figure 4** Performs flowchart for the SSIA.

```
Initialize the salp population xi (i = 1, 2, ..., n) considering ub and lb
while (end condition is not satisfied)
Calculate the fitness of each search agent(salp)
F=the best search agent
SSIA c1 calculated by Eq. (2)
for each salp (xi)
if(i==1)
SSIA the position of the leading salp calculation by Eq. (3)
else
Modify the position of the follower salp by Eq. (4)
end
end
Amend the salps based on the upper and lower bounds of variables
end
return F
```

Figure 3.
Pseudo code of the SSIA algorithm [12].

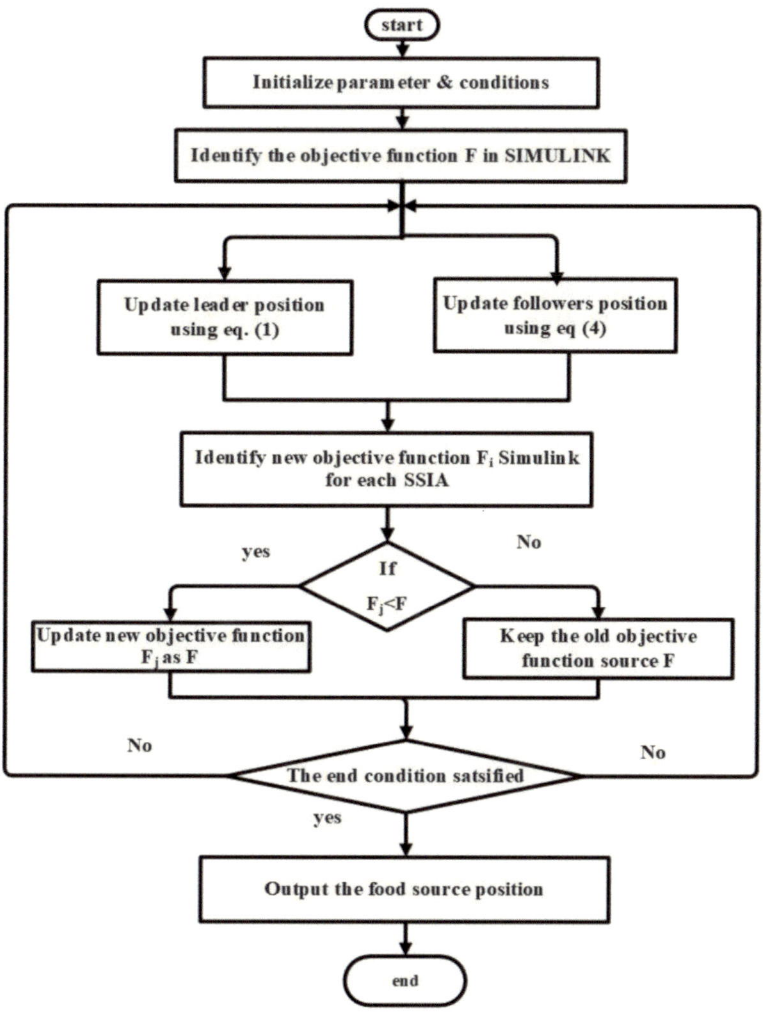

Figure 4.
The flowchart of Salp swarm inspired algorithm.

3. Concept droop control

The ability of the inverter device is one of the fundamental objectives on which the wireless structure relies to regulate the output voltage and frequency while sharing the power and reactive demands. A key to wireless techniques is the use of droop control [33]. This is widely used in conventional power generation systems. A feature of this is that an external communication mechanism between inverters is not needed. In addition, its simple structure, based solely on the local voltage and current data, makes it possible for plug-and-play operations. The importance of droop control power in the island mode becomes obvious when it is possible to provide energy sharing across all units without the need to communicate with other units [3]. This method is built on the droop control of synchronous generators. The active and reactive power of each DG is determined with its nominal capacity and the droop coefficient. This is achieved by changing the droop coefficient, which increases the output resistance of the DG inverters, to regulate the amount of energy injected per DG on the grid. The output voltage and frequency of the inverter is controlled based on the reference active and reactive power of DGs, so the Q-V and P-f droop controllers are usually good candidates [34]. Therefore, the active power will be controlled according to the phase angle, whereas the voltage difference will regulate the reactive power. **Figure 5** demonstrates the relationship between P-ω and Q-V.

In the case of islands, the essential feature of droop control is to control the output power to achieve good power-sharing between transformers. This topology of the three-layer microgrid control strategy with its components will be addressed in detail in the following subsections.

3.1 Power circuit

In the power circuit, four components are used the three-phase VSI, the filter for resistive-inductive-capacitive (RLC), the inductor coupling (L_2), and the three-phase load.

3.2 Droop control

Droop control is a control technique that is usually applied to generators to allow parallel generators to be controlled by the microgrid. The relationship is

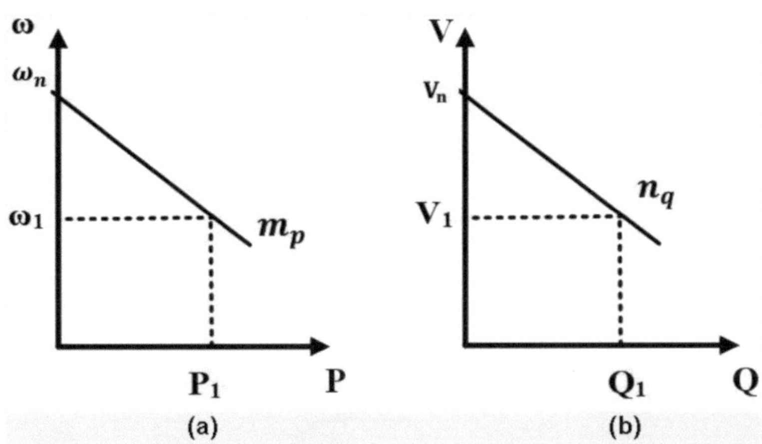

Figure 5.
P-ω and Q-V droop control curves [34].

concentrated between the active power and the frequency and the reactive power and the voltage. Using the output voltage (ν_0) and output current (I_0) to calculate the active power (p) and the reactive power (q) before the filter, V0 and I0 are transformed to the dq reference frame for the calculation of (p and q) using below Equation [8, 35]:

$$p = \nu_{0d} I_{0d} + \nu_{0q} I_{0q} \tag{5}$$

$$q = \nu_{0d} I_{0q} + \nu_{0q} I_{0d} \tag{6}$$

where
ν_{od}: the output voltage on the d reference frame.
ν_{oq}: the output voltage on the q reference frame.
I_{0d}: the output current on the d reference frame.
I_{0q}: the output current on the q reference frame.
P: the active power before the filter.
Q: the reactive power before the filter.

For enhancement, the p and q pass into a low pass filter and are renamed as P and Q. P and Q are determined in accordance with the following equation:

$$P = \frac{\omega_c}{S + \omega_c} \left(\nu_{0d} I_{0d} + \nu_{0q} I_{0q} \right) \tag{7}$$

$$Q = \frac{\omega_c}{S + \omega_c} \left(\nu_{0d} I_{0q} + \nu_{0q} I_{0d} \right) \tag{8}$$

where
ω_c: the cut-off frequency of low-pass filters.
S: the Laplace transform parameter.
P: the measured active power.
Q: the measured reactive power.

After calculating P and Q, the reference angular frequency ω and reference voltage V will be calculated using the equation below:

$$\omega = \omega_n - m_p * P \tag{9}$$

$$V = V_n - n_q * Q \tag{10}$$

where
ω: the reference angular frequency.
V: the reference voltage.
ω_n: the constant coefficients of frequency characteristics.
V_n: the constant coefficients of voltage characteristics.
m_p and n_q: the droop coefficients.

3.3 Voltage: current controller

An input to the voltage controller to find the reference current (I_i^*) will be the reference voltage and frequency. The voltage controller output I_i^* will feed the current controller. The current controller output (V^*) feeds the Pulse Width Modulation (PWM). The output of PWM is used to regulate VSI. I_i^* and V^* are determined by the equations below [8]:

$$I_i^* = -\omega C_f V_o^* + k_{pv} \left(V_o^* - V_o \right) + \frac{K_{iv}}{s} \left(V_o^* - V_o \right) \tag{11}$$

$$V^* = -\omega L_f I_i + k_{pc}\left(I_i^* - I_i\right) + \frac{k_{ic}}{S}\left(I_i^* - I_i\right) \tag{12}$$

where

$L_f$is the coupling inductor.

S is the Laplace transform parameter.

The voltage and current are controlled through the use of the PI controller. The gains of the PI controller and droop coefficients need to be exactly calculated. To determine PI gains and droop coefficients, there are many methods used, such as the trial and error method and the root locus method. However, these methods do not deal with the complicated nonlinear framework, such as microgrids, or even assess the controller's exact gains. Several studies are attempting to solve this issue because of the relevance of calculating PI gains and drop coefficients. So, the SSIA will be applied to obtain PI gains and drop coefficients.

4. SSIA application in microgrids

The proposed SSIA technology will be used to determine the optimum control parameters and drop control coefficients. SSIA determines the control parameters and droop coefficients (K_{p1}, K_{i1}, K_{p2}, K_{i2}, K_{p3}, K_{i3}, K_{p4}, K_{i4}, n_q, m_p) for the realization of minimized voltage and frequency fluctuations. Every optimization technique requires an objective function to perform its assigned task.

The objective function is designed to minimize the error between the calculated and expected voltage. Integral of absolute error (IAE), integral of square error (ISE), integral of time absolute error (ITAE), integral of time square error (ITSE) are the four types of error benchmark objective functions [36]. ITAE is the most widely used feature in literature for reducing control objectives. This is because ITAE aims for easier implementation and provides improved efficiency compared to its rivals. The ITSE and ISE are aggressive criteria and produce unrealistic assessments due to squaring of the mistake made. In contrast to the ITAE, the IAE is also an ineffective choice, reflecting reasonable a more practical error-index due to the time-multiplying error feature. ITAE mathematically explains the equation below:

$$\text{ITAE} = \int_0^\infty t.|e(t)|.dt \tag{13}$$

where

ITAE: integral of time absolute error.

t: time.

e: error.

The multi-objective function is used in this case study to recognize both the frequency and voltage errors via the property of the accumulative sum. **Figure 6** illustration the test system diagram consists of two solar PV array systems (SPVAS), a DC-DC boost converter, two battery stations (BSs), a supercapacitor (SC), a three-phase VSI, a load, and a transmission line [31]. Owing to their fast charging and discharging characteristics, supercapacitors are also used to boost the microgrid's dynamic response. The DC-DC boost converter is fitted with maximum power point tracking (MPPT) based on incremental conductance (INC) to control the DC voltage of the SPVAS output terminals. **Table 1** reviews the parameters of the test system (islanded microgrid model) [8]. The detailed comparative analysis is given in **Table 2** for three optimization techniques. The findings indicate that SSIA

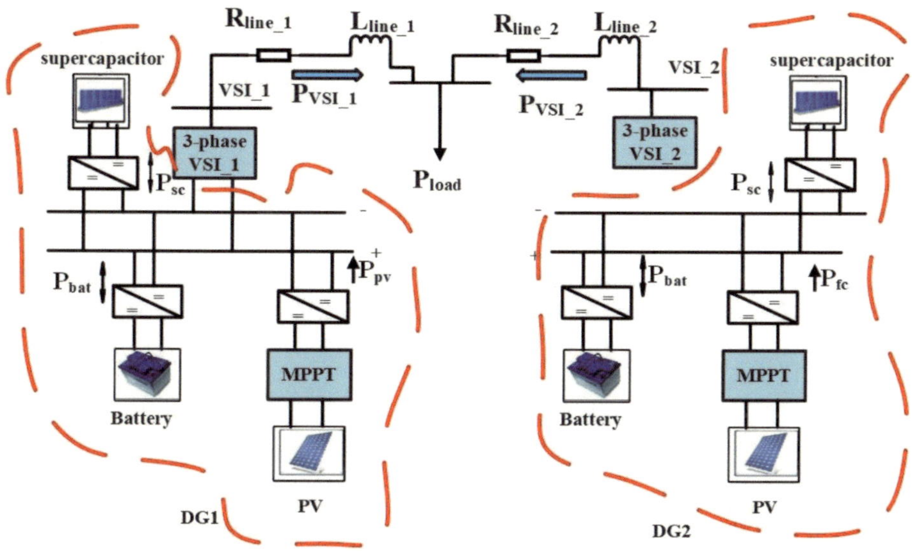

Figure 6.
Test system diagram [31].

Parameter	Value	Parameter	Value
V_{base}	380 V	ω_n	1 p.u.
S_{base}	100 kVA	V_n	1 p.u.
ω_{base}	314 red/sec	R_{line1}	0.14 p.u.
L_f	0.95 x10^{-3} p.u.	L_{line1}	2.1 x10^{-3} p.u.
C_f	35 x10^{-6} p.u.	R_{line2}	0.2 p.u.
R_f	0.067 p.u.	L_{line2}	3.5 x10^{-3} p.u.
L_c	0.23 x10^{-3} p.u.	P_{load}	70 x10^3
R_c	0.02 p.u.	ω_c	0.1 p.u.
Ts Power of PV Power of battery	5.144 x10^{-6} sec 109.88 kW 56 kW	Frequency of PWM Capacitance of supercapacitor	10 kHz 29 F

Table 1.
Test system parameters [8].

	SSA	PSO	ABC
Objective function	10.25 x10^6	12.66 x10^6	18.460 x10^6
K_{p1}	0.691433	0.557908	0.641502
K_{i1}	311.4964	475.0824	530.0400
K_{p2}	0.548548	0.513765	0.470766
K_{i2}	543.6527	565.1678	349.2290
K_{p3}	11.27808	13.09909	12.22560
K_{i3}	13879.68	12410.52	6358.140
K_{p4}	10.25541	13.05646	9.003440
K_{i4}	14414.39	11132.63	13267.80

	SSA	PSO	ABC
n_q	0.272247	0.225909	0.266158
m_p	0.014868	0.009724	0.015636
Time Taken (min)	207.2686	219.1358	224.4803

Table 2.
Results of the applied three optimization techniques.

succeeded with minimal voltage and frequency errors in achieving the assigned control task. Three alternative optimization methods (SSIA, Particle Swarm Optimization (PSO), and ABC) are used to validate the quality of SSIA for a fair comparison.

5. Results of simulations

On a microgrid test device, well-tuned controllers via SSIA are equipped to confirm the power-sharing between multiple sources as well as the voltage and frequency regulation. The two types of load: constant and continuous change loads are considered in this study with RERs variability such as (variable irradiance and temperature).

5.1 Case I: islanding mode with fixed cyclic load variations scenario (IMFCLVS)

In this case, for islanded MG with a 70 kW (0.7p.u.) constant load, RERs variability (variable solar irradiance and temperature) is regarded. The ramp-up/

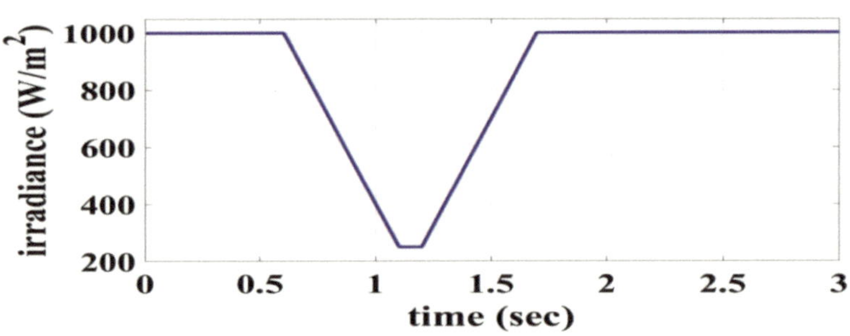

Figure 7.
Solar irradiance variation pattern for all applied scenarios.

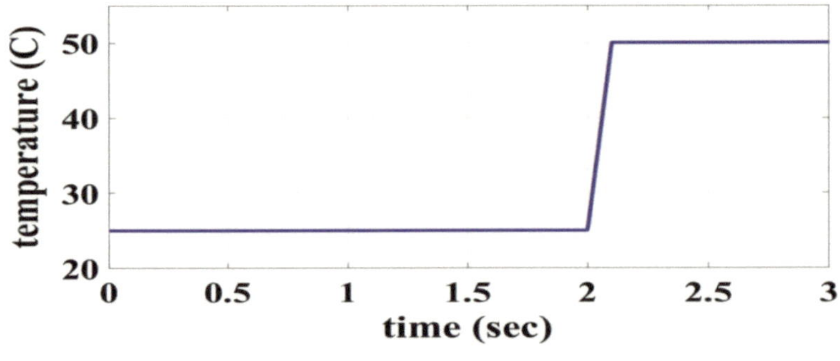

Figure 8.
Solar temperature variation for all applied scenarios.

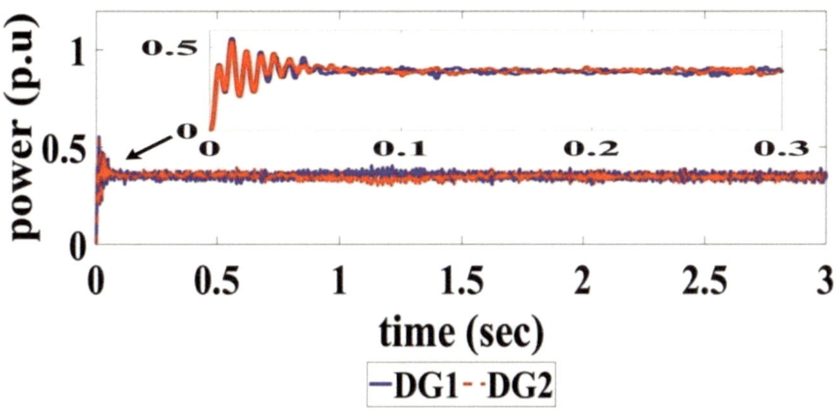

Figure 9.
Active powers generated by two DGs IMFCLVS scenario.

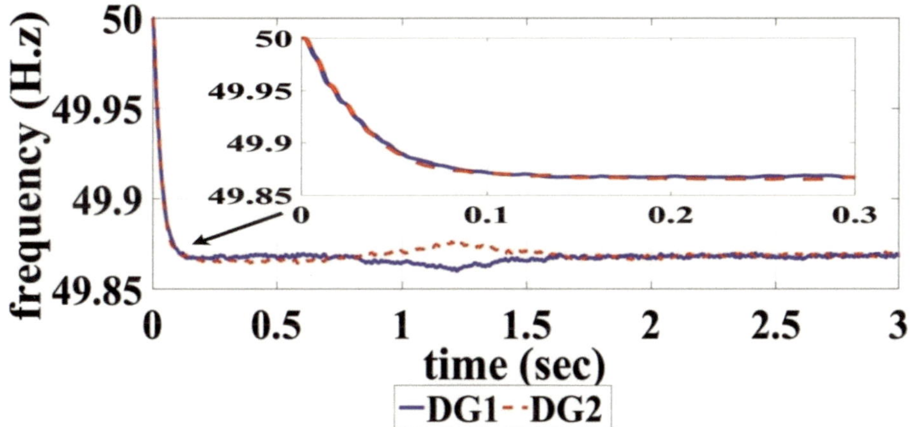

Figure 10.
Inverter frequency of IMFCLVS scenario.

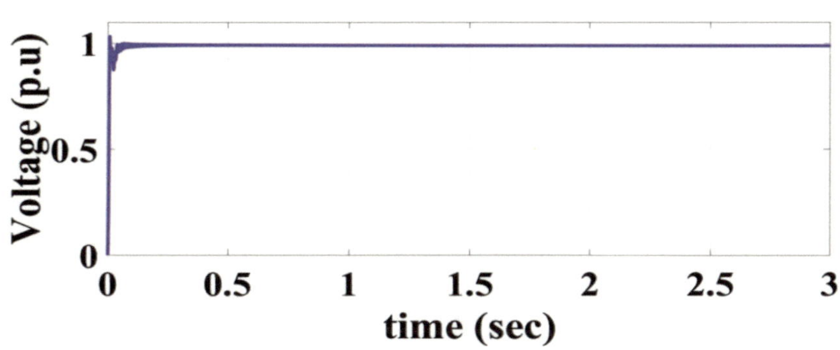

Figure 11.
Voltage magnitude of IMFCLVS scenario.

down solar irradiance from 1000 W/m2 to 250 W/m2 is expressed in **Figure** 7. **Figure 8** establishes the temperature variation between 25°C and 50°C. For each source, the dynamic active power response is represented in **Figure 9**. It should be noted that for both sources, the active power is almost equal (0.35 p.u.), which confirms successful power-sharing. Remarkably, it is observed that solar radiation

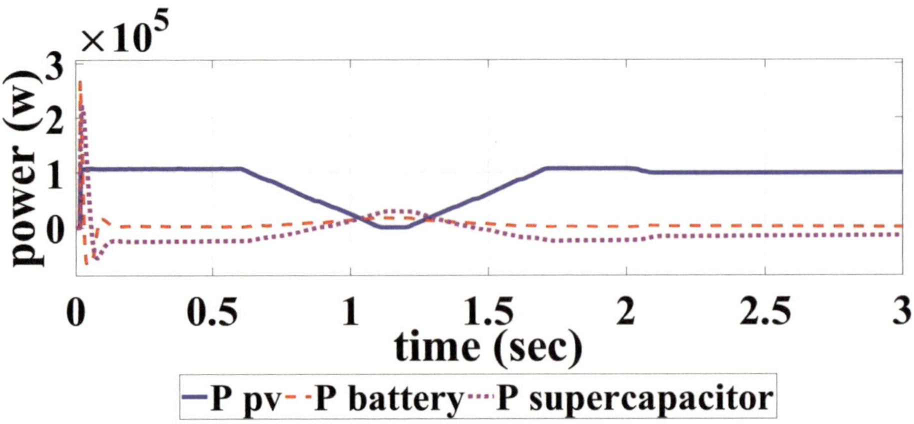

Figure 12.
Powers of SPVASs, BSs, and SC in IMFCLVS scenario.

and temperature fluctuations do not impact power-sharing because the total power fed to the DC bus is constant due to the mechanism of energy management. **Figures 10** and **11** indicate the frequency and voltage responses during the applied case. **Figure 12** shows the process of energy management among the multi-sources of MG (SPVAS, BS, and SC). It is clear from the findings that the droop control strategy based on SSIA dealt successfully with the variability of RERs.

5.2 Case II: islanding mode with continuous cyclic load variations scenario (IMCCLVS)

The microgrid is operated in islanding mode with continuous cyclic load variations under the variability of RERs as kW (0.7 p.u.) from 0 to 0.3 sec, then the load value increased to 110 kW (1.1 p.u.) at 0.3–0.7 sec, then the load value returned to 70 kW (0.7 p.u.) at 0.7–1.2 sec at the end of the load cycle. **Figure 13** shows that an equivalent amount of active power is injected into MG by each DG. The rate of power change is notably almost the same as the rate of load change, remarkably. **Figures 14** and **15** denote the transient response of frequency and voltage during the IMCCLVS scenario, respectively. Strikingly, **Figures 14** and **15**. show that the frequency response indicates that the rate of power change is highly influenced by the frequency response, while the voltage is slightly affected. Additionally,

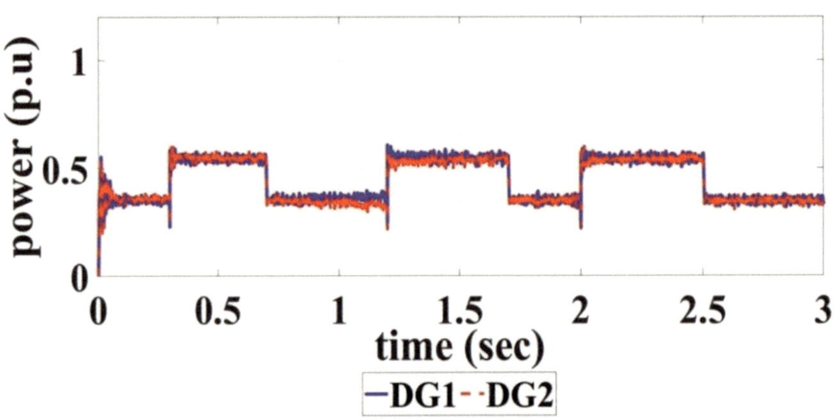

Figure 13.
Active powers generated by DGs in IMCCLVS scenario.

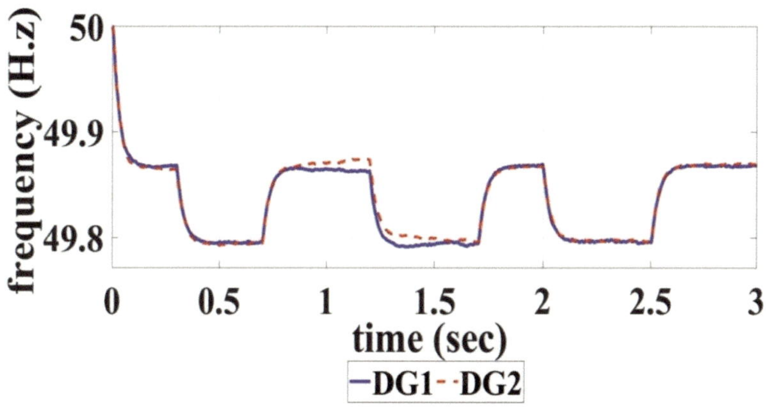

Figure 14.
Inverter frequency of IMCCLVS scenario.

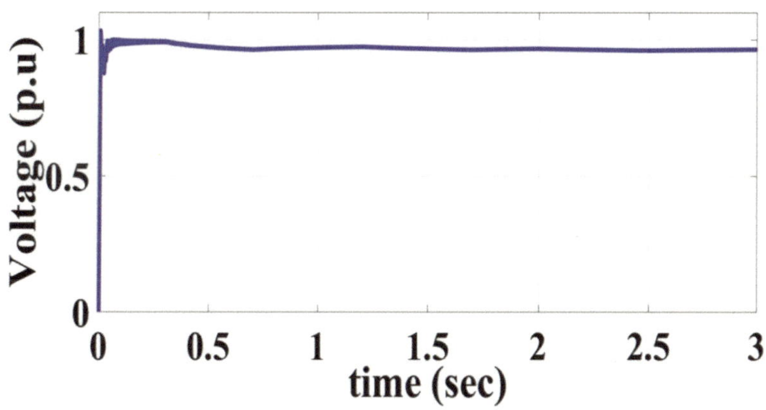

Figure 15.
Voltage magnitude IMCCLVS scenario.

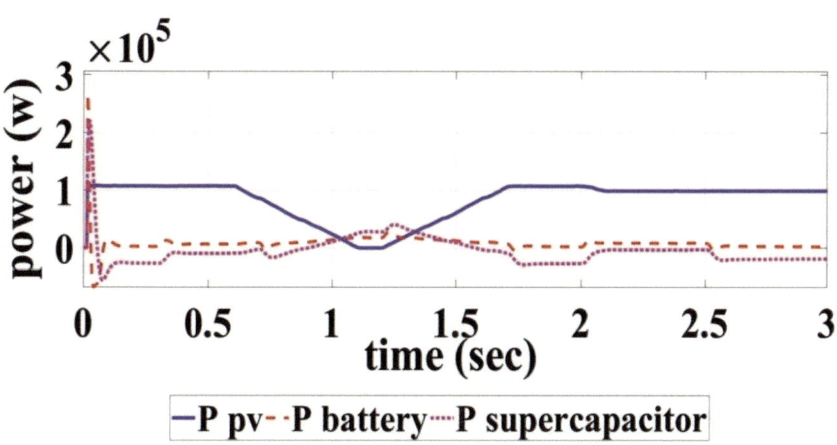

Figure 16.
Powers of SPVASs, BSs, and SC in IMCCLVS scenario.

Figure 16 demonstrates how SPVASs, BSs, and SC interact in dynamism with each other to preserve the continuity of supply.

6. Conclusions

In this chapter, an optimal SSIA-based voltage and frequency control and the power-sharing scheme was presented for inverter-dependent DG units in an island microgrid. Two sources and each source of a microgrid test system are available. The solar PV array, supercapacitor and battery station are included. To determine the gains of the PI controllers and coefficients of the droop control system, the SSIA is used. The cost function includes four forms: IAE, ISE, ITAE, and ITSE. The best solution is found when implementing ITAE as an objective function. By comparing three different types of optimization techniques (SSIA, PSO and ABC) applied to the microgrid scheme, the quality of SSIA as an optimization method was verified. In two cases, the achieved gains of PI controllers and droop control coefficients (K_{p1}, K_{i1}, K_{p2}, K_{i2}, K_{p3}, K_{i3}, K_{p4}, K_{i4}, n_q, m_p) are implemented in the system. Fixed load and slow and fast changes are considered in the proposed cases, as well as abrupt variations in both renewable energy supplies and loads. The results of the simulation indicated that the power-sharing depends on SSIA between the parallel DGs and the droop control strategy. The frequency deviation is within the acceptable range and the DGs are easily followed by changes in load with a good dynamic response.

Conflict of interest

The authors have no affiliation with any organization with a direct or indirect financial interest in the subject matter discussed in the manuscript.

Author details

Mohamed A. Ebrahim[1*], Reham M. Abdel Fattah[2], Ebtisam M. Saied[1,3], Samir M. Abdel Maksoud[1] and Hisham El Khashab[2]

1 Electrical Engineering Department, Faculty of Engineering at Shoubra, Benha University, Cairo, Egypt

2 Power Electronics and Energy Conversion Department, Electronics Research Institute, Cairo, Egypt

3 Electrical Engineering Department-High Technological Institute (HTI) 10th of Ramadan City, Egypt

*Address all correspondence to: mohamed.mohamed@feng.bu.edu.eg

References

[1] Jumani TA, Mustafa MW, Rasid MM, Mirjat NH, Leghari ZH, Salman Saeed M. Optimal voltage and frequency control of an islanded microgrid using grasshopper optimization algorithm. Energies 2018;11. https://doi.org/10.3390/en11113191.

[2] Khaledian A, Golkar MA. Analysis of droop control method in an autonomous microgrid 2017;15:371–7.

[3] Tayab UB, Roslan MA Bin, Hwai LJ, Kashif M. A review of droop control techniques for microgrid. Renew Sustain Energy Rev 2017;76:717–27. https://doi.org/10.1016/j.rser.2017.03.028.

[4] Goodarzi HM, Kazemi MH, Kazemi MH. An optimal autonomous microgrid cluster based on distributed generation droop parameter optimization and renewable energy sources using an improved grey wolf optimizer. Eng Optim 2017;0:1–21. h ttps://doi.org/10.1080/0305215X.2017.1355970.

[5] Zeng Z, Yang H, Zhao R. Study on small signal stability of microgrids: A review and a new approach. Renew Sustain Energy Rev 2011;15:4818–28. https://doi.org/10.1016/j.rser.2011.07.069.

[6] Shokoohi S, Bevrani H. An Intelligent Droop Control for Simultaneous Voltage and Frequency Regulation in Islanded Microgrids. IEEE Trans Smart Grid 2013;4:1505–13. https://doi.org/10.1016/j.ijepes.2014.07.024.

[7] Zeng X, Tang C, Wang W, Tang X. Analysis of microgrid inverter droop controller with virtual output impedance under non-linear load condition. IET Power Electron 2014;7:1547–56. https://doi.org/10.1049/iet-pel.2013.0407.

[8] Kohansal M, Gharehpetian GB, Abedi M. An optimization to improve voltage response of VSI in islanded Microgrid considering reactive power sharing. 2012 2nd Iran Conf Renew Energy Distrib Gener ICREDG 2012 2012:127–31. https://doi.org/10.1109/ICREDG.2012.6190447.

[9] Aljarah I. Grasshopper optimization algorithm for multi-objective optimization problems 2017. https://doi.org/10.1007/s10489-017-1019-8.

[10] Tsai C, Eberle W, Chu C. Knowledge-Based Systems Genetic algorithms in feature and instance selection. Knowledge-Based Syst 2013;39:240–7. https://doi.org/10.1016/j.knosys.2012.11.005.

[11] Heidari AA, Mirjalili S, Faris H, Aljarah I, Mafarja M, Chen H. Harris hawks optimization: Algorithm and applications. Futur Gener Comput Syst 2019;97:849–72. https://doi.org/10.1016/j.future.2019.02.028.

[12] Mirjalili S, Gandomi AH, Mirjalili SZ, Saremi S, Faris H, Mirjalili SM. Salp Swarm Algorithm: A bio-inspired optimizer for engineering design problems. Adv Eng Softw 2017;114:163–91. https://doi.org/10.1016/j.advengsoft.2017.07.002.

[13] Abbassi R, Abbassi A, Asghar A, Mirjalili S. An efficient salp swarm-inspired algorithm for parameters identification of photovoltaic cell models An efficient salp swarm-inspired algorithm for parameters identification of photovoltaic cell models. Energy Convers Manag 2018;179:362–72. https://doi.org/10.1016/j.enconman.2018.10.069.

[14] Hegazy AE, Makhlouf MA, El-Tawel GS. Improved salp swarm algorithm for feature selection. J King Saud Univ - Comput Inf Sci 2018. https://doi.org/10.1016/j.jksuci.2018.06.003.

[15] Saremi S, Mirjalili S, Lewis A. Grasshopper Optimisation Algorithm: Theory and application. Adv Eng Softw 2017;105:30–47. https://doi.org/10.1016/j.advengsoft.2017.01.004.

[16] Mirjalili S, Lewis A. The Whale Optimization Algorithm. Adv Eng Softw 2016;95:51–67. https://doi.org/10.1016/j.advengsoft.2016.01.008.

[17] Ebrahim MA, Osama A, Kotb KM, Bendary F. Whale inspired algorithm based MPPT controllers for grid-connected solar photovoltaic system. Energy Procedia 2019;162:77–86. https://doi.org/10.1016/j.egypro.2019.04.009.

[18] Mafarja MM, Mirjalili S. Hybrid Whale Optimization Algorithm with simulated annealing for feature selection. Neurocomputing 2017;260:302–12. https://doi.org/10.1016/j.neucom.2017.04.053.

[19] Mirjalili S. Moth-flame optimization algorithm: A novel nature-inspired heuristic paradigm. Knowledge-Based Syst 2015;89:228–49. https://doi.org/10.1016/j.knosys.2015.07.006.

[20] Mirjalili S, Mirjalili SM, Lewis A. grey wolf. Adv Eng Softw 2014;69:46–61. https://doi.org/10.1016/j.advengsoft.2013.12.007.

[21] Ahmed Ebrahim M, Mohamed RG. Comparative Study and Simulation of Different Maximum Power Point Tracking (MPPT) Techniques Using Fractional Control and Grey Wolf Optimizer for Grid Connected PV System with Battery. Electr Power Convers 2019;2016:1–14. https://doi.org/10.5772/intechopen.82302.

[22] Soued S, Ebrahim MA, Ramadan HS, Becherif M. Optimal blade pitch control for enhancing the dynamic performance of wind power plants via metaheuristic optimisers. IET Electr Power Appl 2017;11:1432–40. https://doi.org/10.1049/iet-epa.2017.0214.

[23] Mirjalili S. The ant lion optimizer. Adv Eng Softw 2015;83:80–98. https://doi.org/10.1016/j.advengsoft.2015.01.010.

[24] Aouchiche N, Aitcheikh MS, Becherif M, Ebrahim MA. AI-based global MPPT for partial shaded grid connected PV plant via MFO approach. Sol Energy 2018;171:593–603. https://doi.org/10.1016/j.solener.2018.06.109.

[25] Ebrahim MA, Becherif M, Abdelaziz AY. Dynamic performance enhancement for wind energy conversion system using Moth-Flame Optimization based blade pitch controller. Sustain Energy Technol Assessments 2018;27:206–12. https://doi.org/10.1016/j.seta.2018.04.012.

[26] Ali AM, Ebrahim MA, Moustafa Hassan MA. Automatic voltage generation control for two area power system based on particle swarm optimization. Indones J Electr Eng Comput Sci 2016;2:132–44. https://doi.org/10.11591/ijeecs.v2.i1.pp132-144.

[27] Ebrahim MA, Mostafa HE, Gawish SA, Bendary FM. Design of decentralized load frequency based-PID controller using stochastic particle swarm optimization technique. 2009 Int Conf Electr Power Energy Convers Syst EPECS 2009 2009.

[28] Mirjalili S. Dragonfly algorithm: a new meta-heuristic optimization technique for solving single-objective, discrete, and multi-objective problems. Neural Comput Appl 2016;27:1053–73. https://doi.org/10.1007/s00521-015-1920-1.

[29] Mirjalili S, Gandomi AH, Mirjalili SZ, Saremi S, Faris H, Mirjalili SM. Salp Swarm Algorithm: A bio-inspired optimizer for engineering design problems. Adv Eng Softw 2017;114:163–91. https://doi.org/10.1016/j.advengsoft.2017.07.002.

[30] Abbassi R, Abbassi A, Heidari AA, Mirjalili S. An efficient salp

swarm-inspired algorithm for parameters identification of photovoltaic cell models. Energy Convers Manag 2019;179:362–72. https://doi.org/10.1016/j.enconman.2018.10.069.

[31] Ebrahim MA, Fattah RMA, Saied EM, Maksoud SMA, El Khashab H. Real-Time Implementation of Self-Adaptive Salp Swarm Optimization-based Microgrid Droop Control. IEEE Access 2020;8:1–1. https://doi.org/10.1109/access.2020.3030160.

[32] Qais MH, Hasanien HM, Alghuwainem S. Enhanced salp swarm algorithm: Application to variable speed wind generators. Eng Appl Artif Intell 2019;80:82–96. https://doi.org/10.1016/j.engappai.2019.01.011.

[33] Lopes JAP, Moreira CL, Madureira AG. Defining control strategies for microgrids islanded operation. IEEE Trans Power Syst 2006;21:916–24. https://doi.org/10.1109/TPWRS.2006.873018.

[34] Zafari P, Zangeneh A, Moradzadeh M, Ghafouri A, Parazdeh MA. Various Droop Control Strategies in Microgrids. Power Syst 2020:527–54. https://doi.org/10.1007/978-3-030-23723-3_22.

[35] Ebrahim MA, Aziz BA, Nashed MNF, Osman FA. A Novel Hybrid-HHOPSO Algorithm based Optimal Compensators of Four-Layer Cascaded Control for a New Structurally Modified AC Microgrid. IEEE Access 2020. https://doi.org/10.1109/ACCESS.2020.3047876.

[36] Hussain KM, Zephcrin RAR, Kumar MS, Kumar SMG. Comparison of PID Controller Tuning Methods with Genetic Algorithm for FOPTD System. J Eng Res Appl 2014;4:308–14.

Section 6

Estimation of Hidden Energy Losses in Technological Systems

Chapter 6

Estimation of Hidden Energy Losses

Borys Pleskach

Abstract

Most industrial or municipal energy consumers involve the conversion of electricity, either into useful products or into other types of energy. For example, lighting systems, heating systems, air or water supply systems. And in all such systems there are energy losses, which can be divided into open, or technological and hidden, or abnormal. Open losses are inherent in the technological process itself and depend on the principle of energy conversion, flow conditions, the type of equipment received, and so on. Hidden losses in the technological system occur accidentally due to the appearance of defects in the equipment, erroneous actions of personnel, changes in uncontrolled external conditions. The paper considers a method of detecting and estimating hidden energy losses, based on the analysis of energy consumption precedents and building a decision support system aimed at eliminating such energy losses. Models of energy consumption precedents are formed on the basis of controlled technological parameters and their statistical estimates. In the future, local standards of efficient energy consumption are formed from individual precedents. The advantage of this method of estimating latent energy losses is the adaptation of standards of efficient energy consumption to the conditions of the consumer.

Keywords: Energy losses, energy efficiency, energy saving, precedent analysis, energy consumers

1. Introduction

The increase in the cost of energy and their availability, as well as the growing impact of energy-consuming technological systems on the environment, require the introduction of systematic measures aimed at improving the energy efficiency of production. At the same time, it is predicted [1] that energy consumption will increase in the future. This will especially affect electricity. Worldwide, about 50% of total electricity consumption is carried out in industry by conversion using electric motor systems [2]. The highly automated industry of developed countries converts almost 60-75% of all electricity distributed in four types of electric motor systems into mechanical movement of about 40%, of which compressors 25%, pumps 20% and fans 15%. This study shows that the control system, based on the analysis of technological parameters and intensity of electricity consumption from electric motor systems, can be widely used in industry [3].

The gradual transition from traditional to renewable energy sources requires an improvement in the balance of supply and demand, the creation of active energy consumers capable of controlling their own energy efficiency and maintaining their own energy sources. Eliminating hidden energy losses in both energy consumption and generation gives businesses additional benefits.

Improving the energy efficiency of consumers in industry was mainly achieved through the introduction of new technologies, new more economical equipment, construction using new energy-saving materials, heat recovery, and so on. To a lesser extent, organizational approaches to energy consumption management were introduced, i.e. planning and production management taking into account the economically feasible use of energy resources. This is due to the fact that organizational approaches are closely integrated into production planning and control systems, and require additional investment costs for the development of information technology aimed at energy saving. Obtaining information on energy consumption and linking it to operational information to get an idea of the behavior of equipment and technological systems has proven to be technically difficult. However, the potential for energy savings from organizational approaches is quite significant [4]. Energy management combines the overall management of the enterprise with the needs of improving the energy efficiency of technological systems.

This work is aimed at the study of operational monitoring and analysis of energy consumption as a sequence of cases of stationary and quasi-stationary energy consumption. With regard to energy saving, real-time information can raise awareness of employees of the presence of hidden energy losses and influence their actions. Operational monitoring of deviations of energy consumption from the accepted values of efficient energy consumption can be an indicator of equipment performance and a tool to support maintenance measures. Specific equipment failures can be predicted by identifying certain patterns of energy consumption. Responding quickly to such events can prevent catastrophic equipment failures. Thus, energy monitoring becomes a driver of equipment reliability, technological process stability and product quality.

2. Purpose of work

Most energy consumers convert the received energy and raw materials into a useful product. Technological equipment, production products and performers are involved in this transformation process. Each of these links can affect energy consumption and energy losses. The specific energy consumption of the technological system E is characterized by the vector of influential technological parameters $X = \{x_1, x_2, \dots, x_k\}$ and a set of random uncontrolled factors Z that cause hidden energy losses ΔE.

$$E = f(X, Z) = \varphi(X) + \Delta E \tag{1}$$

Hidden energy losses from the consumer can be caused by violations of the technical condition of the equipment, deterioration of the properties of raw materials, erroneous actions of personnel, and other factors that are usually not subject to automated control. The problem is to provide real-time staff with information on the presence and extent of hidden energy losses in each technological system of the enterprise, based solely on influential technological parameters and specific energy consumption. The purpose of this work is to develop methods for estimating latent energy losses based on the analysis of cases of quasi-stationary energy consumption.

3. Known methods of monitoring energy losses

Energy efficiency of the consumer is determined by the ratio of the amount of energy consumed to the amount of product received, or services provided when working in a certain mode. In the absence of energy losses, this figure will take the minimum values in all modes of operation. When energy losses occur, it will increase. At the same time, the performance of the technological system depends on the parameters that determine its mode of operation, and accordingly on the intensity of energy consumption.

The dependence of energy consumption on the set of technological parameters X in the absence of random uncontrolled factors Z will be called the function of efficient energy consumption, and the value of this function at certain technological parameters X - efficient energy consumption E_{ef}. That is, energy consumption without hidden energy losses:

$$E_{ef} = \varphi\,(X) = f\,(X, Z = 0) \tag{2}$$

Energy efficiency monitoring and, accordingly, the estimation of latent energy losses are based on the comparison of the current energy consumption E with the efficient energy consumption E_{ef} obtained from Eq. (2).

$$\Delta E = E\text{–}E_{ef} \tag{3}$$

The function of efficient energy consumption, and consequently the value of efficient energy consumption, at certain technological parameters, is usually determined in several ways. There is a method when efficient energy consumption is obtained by calculating the known empirical or analytical dependences of electrical engineering, heat engineering, mechanics, hydraulics [5]. This method gives an approximate estimate of the reference energy in the real technological process. Another way to obtain an estimate of efficient energy consumption is based on testing tests of equipment and determining the normative energy consumption when operating equipment under specified conditions [6]. Such estimates of regulatory energy consumption do not always correspond to effective standards in real production [7]. There is also a method when energy efficiency is considered to be achieved a certain period of time ago, when the technological parameters were similar. In most cases, energy management uses a linear regression model of the dependence of “standard” energy consumption on controlled technological parameters according to the “Monitoring and Targeting” method [8] for comparison with current energy consumption.

4. Assessment of hidden energy losses

The paper proposes to use the methods of precedent analysis [9], based on cases of efficient energy consumption, to evaluate and monitor the standards of efficient energy consumption. To do this, each energy consumer in the production process must be equipped with a system of sensors of influential technological parameters associated with a programmable controller of the monitoring system, whose task is to recognize the precedents of quasi-stationary energy consumption.

The precedent of quasi-stationary energy consumption *CaseE* will be the case in which all normalized influential technological parameters $X = \{X_1, X_2, \dots, X_n\}$ for a certain period of time remain within predetermined allowable values $\pm \Delta x_n = \sigma$. The precedent has the following structure:

$$CaseE = \left\langle \begin{array}{ccc} M(X_1), & \dots, & M(X_n); \\ D(X_1), & \dots, & D(X_n); \\ r(X_1), & \dots, & r(X_n); \\ E, & \tau, & S \end{array} \right\rangle \tag{4}$$

where: $M(X_1), \ldots, M(X_n)$ - mathematical expectations of the factors of influence $X_1, \ldots, X_n$;

$D(X_1), \ldots, D(X_n)$ - statistical variance of influencing factors $X_1, \ldots, X_n$;

$r(X_1), \ldots, r(X_n)$ - autocorrelation coefficients of influencing factors $X_1, \ldots, X_n$;

n - is the number of interdependent influential technological parameters;

E - specific energy consumption for the period of quasi-stationary state;

τ - duration of steady state;

S - probable diagnosis of technical condition.

The precedents obtained in this way form in the n-dimensional space of influential technological parameters $\{X_1, X_2, \ldots, X_n\}$ a cloud of precedents of quasi-stationary states of the technological system with different estimates of energy consumption. In this cloud, precedents with minimal energy consumption form the surface of cases of efficient energy consumption.

$$E_{ef} = \varphi(X_1, X_2, \ldots, X_n) = f(X, Z = 0) \tag{5}$$

The estimation of latent energy losses in an arbitrary i-th case (E_i) is based on the determination of a local standard of efficient energy consumption for the i-th precedent. To do this, from the base of precedents of efficient consumption is selected n precedents closest to the current precedent and on them, by the method of least squares, calculate the regression coefficients $b_0, b_1, b_2, \ldots, b_n$ – the function of efficient energy consumption for the current precedent. The value of effective consumption for the current i-th precedent is calculated by the formula:

$$E_{efi} = b_0 + b_1 M(X_{1i}) + b_2 M(X_{2i}) + \ldots + b_n M(X_{ni}) \tag{6}$$

After that, the difference between the obtained value of E_{efi} and the current specific energy consumption E_i is calculated: $\Delta E_i = E_{efi} - E_i$. Depending on the obtained value of ΔE_i it is possible to draw a conclusion about the energy efficiency of the equipment that is subject to monitoring. If $\Delta E_i \approx 0$, the equipment is considered to be operating efficiently, if $\Delta E_i > 0$, the equipment is operating with reduced energy consumption and energy savings are equal to ΔE_i, if $\Delta E_i < 0$, the equipment is operating with energy losses up to ΔE_i.

5. Information support for monitoring hidden energy losses

The flow of derived data from the sensors enters the monitoring system in the form of a time series. The difficulty lies in the synchronization and subsequent search for the relationship between the vector of technological parameters and the intensity of energy costs. This complexity can be overcome by segmenting the time series and allocating stationary areas from its composition, which will be considered as precedents for energy consumption. An overview of possible methods of time series segmentation is given in [10].

Time series $T = \{(X_1, E_1); (X_2, E_2); \ldots, (X_m, E_m)\}$; $1 \le m \le M$ is a finite set of M n-dimensional structured sequences of regime parameters $X = [x_1, x_2, \ldots, x_n]$ and the corresponding energy consumption intensity E, marked by timestamps $t_1, \ldots, t_M$. Segmentation of the time series T divides it into a sequence of independent series $T = \{S_1, S_2, \ldots, S_i\}$ that do not overlap. The segment S of the time series T is a finite set of structured elements (X_m, E_m) marked with timestamps $t_a, \ldots, t_n, \ldots, t_b$: $S_i = \{(X_a, E_a); \ldots; (X_b, E_b)\}$.

The purpose of the procedure of segmentation of the time series of mode parameters is to divide the data flow into separate disparate areas with similar characteristics and to allocate among them areas with signs of stationarity [11]. To formalize the relationship between the elements of the series, a special function of the price of entry into the segment *Cost{S}* is introduced, which determines the relationship between the elements of the series. Usually the function of distance between elements of a series or groups of elements is used for segmentation [12, 13]. In this paper, for time series segmentation, it is proposed to use the function of the distance d between the elements X_1 and X_i of the time series in the n-dimensional Euclidean space of mode parameters X:

$$Cost\{S\} = d(X_1, X_i) = \sqrt{(x_{11} - x_{1i})^2 + (x_{21} - x_{2i})^2 + \cdots + (x_{k1} - x_{ki})^2} \quad (7)$$

The condition of inclusion of the next, i-th, group of the received mode parameters X_i, E_i to the next segment S, which is filled with data:

$$Cost\{S\} = d(X_1, X_i) \le \sigma \quad (8)$$

where: σ is a threshold value that is determined empirically.

The pseudocode of the algorithm segmentation of the time series of determining mode parameters is as follows:

- open a time series segment;
- get and save in the open segment the first set of mode parameters X_1, E_1;
- obtain the next set of mode parameters X_i, E_i and calculate the entry price of this set in the open segment according to formula (7).
- as long as the segment entry price for the new sets of mode parameters is less than the threshold value σ, according to formula (8), they join the open segment. Otherwise, the open segment is closed and passed for further processing, a new segment is opened and it is the first to enter a set of mode parameters for which the entry price was higher than the threshold value. And then the cycle repeats.

The algorithm in the pace of the technological process processes the derivative flow of mode parameters, determines the segments with signs of stationarity and transmits them for further processing to calculate the hidden energy losses. The procedure of segmentation of the flow of mode parameters of the technological system can be implemented in a separate programmable controller located in the area of energy metering.

6. Formation of a base of precedents

During the assessment of latent or abnormal energy losses in the composition of the upper-level computing device, a base of precedents of quasi-stationary consumption is formed. The initial filling of the precedent database is performed with the participation of an expert. Let *Case* be a set of precedents. It is believed that all precedents of energy consumption are located in an unlimited space of mathematical expectations of regime factors. In this space the function of distance between precedents is set:

$$\mathrm{d(Casei\ Casej)} = \sqrt{\sum (M(X_{ni}) - M(X_{nj}))^2} \quad (9)$$

Considerations based on the analysis of precedents are to search among all recorded precedents of the nearest neighbors to the current one and to build a linear approximation of the function of efficient energy consumption in the nearest neighbors. The search for the nearest neighbors is carried out among a large number of precedents. To simplify this procedure, it is proposed to involve a clustering mechanism. Clustering of precedents is the process of combining precedents into groups characterized by similar features. Unlike the usual classification, where the number of groups of objects is fixed and predetermined by a set of precedents, here neither groups nor their number are predetermined and formed in the process of the system, based on a certain degree of proximity of precedents [14, 15].

The formed clusters represent separate disjoint areas of function (2). It can be assumed that the precedents assigned to the same cluster belong to one area of the energy efficiency function. These clusters, in turn, are proposed to be used as a basis for determining the proximity of precedents. Suppose that we gradually obtain a sequence of precedents of static energy consumption. It is necessary to assign each of them to one of the disjoint subsets at the rate of obtaining precedents so that each cluster consists of precedents, which according to the metric $d(Case_i\ Case_j)$ were in the area of existence of the cluster. In this case, each precedent is assigned the identifier of the cluster $n \in N$, to which it belongs. We assume that the space of mathematical expectations of influencing factors has zero precedent $Case_0$ with zero values of influencing factors. The distance from the zero precedent to any other precedent is determined by the dependence:

$$d(Case_0, Case_i) = \sqrt{\left(M_i(X_1)^2 + M_i(X_2)^2 + \cdots + M_i(X_n)^2\right)} \quad (10)$$

Each of the clusters is characterized by an initial radius R_i and a depth of existence ΔR. The initial radius of the cluster is the distance between the zero precedent of $Case_0$ and the nearest cluster boundary (**Figure 1**) $R_i = d\ (Case_0, Case_i)$. The area of existence of the cluster is a set of $Case_i$ precedents (marked with asterisks in the figure) that satisfy the condition $R_i \leq d\ (Case_0,\ Case_i) < R_i + \Delta R$.

The clustering algorithm is a function a: $Case_i \rightarrow n$ which matches the cluster identifier $n \in N$ to any precedent, and the set N is unknown in advance. As the

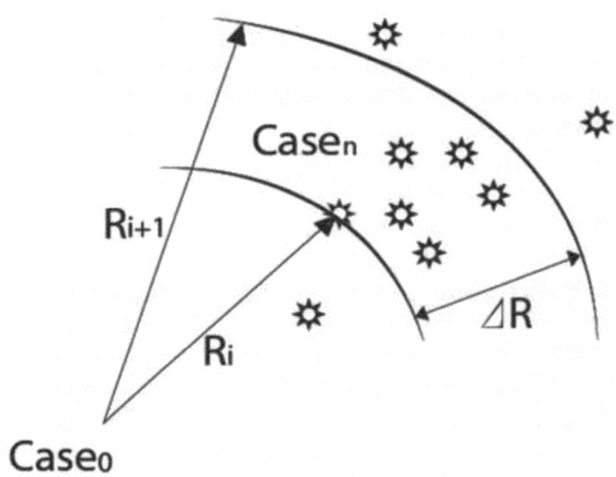

Figure 1.
Cluster diagram.

identifier n of the cluster, it is advisable to use its radius, which is calculated by the formula:

$$n_i = R_i = \left\lfloor \frac{d(Case_0\, Case_j)}{\Delta R} \right\rfloor \tag{11}$$

In the future, we look for the nearest neighbors only among the precedents of the current cluster. New precedents are placed in the base of effective precedents only in the absence of energy losses (see **Figure 2**).

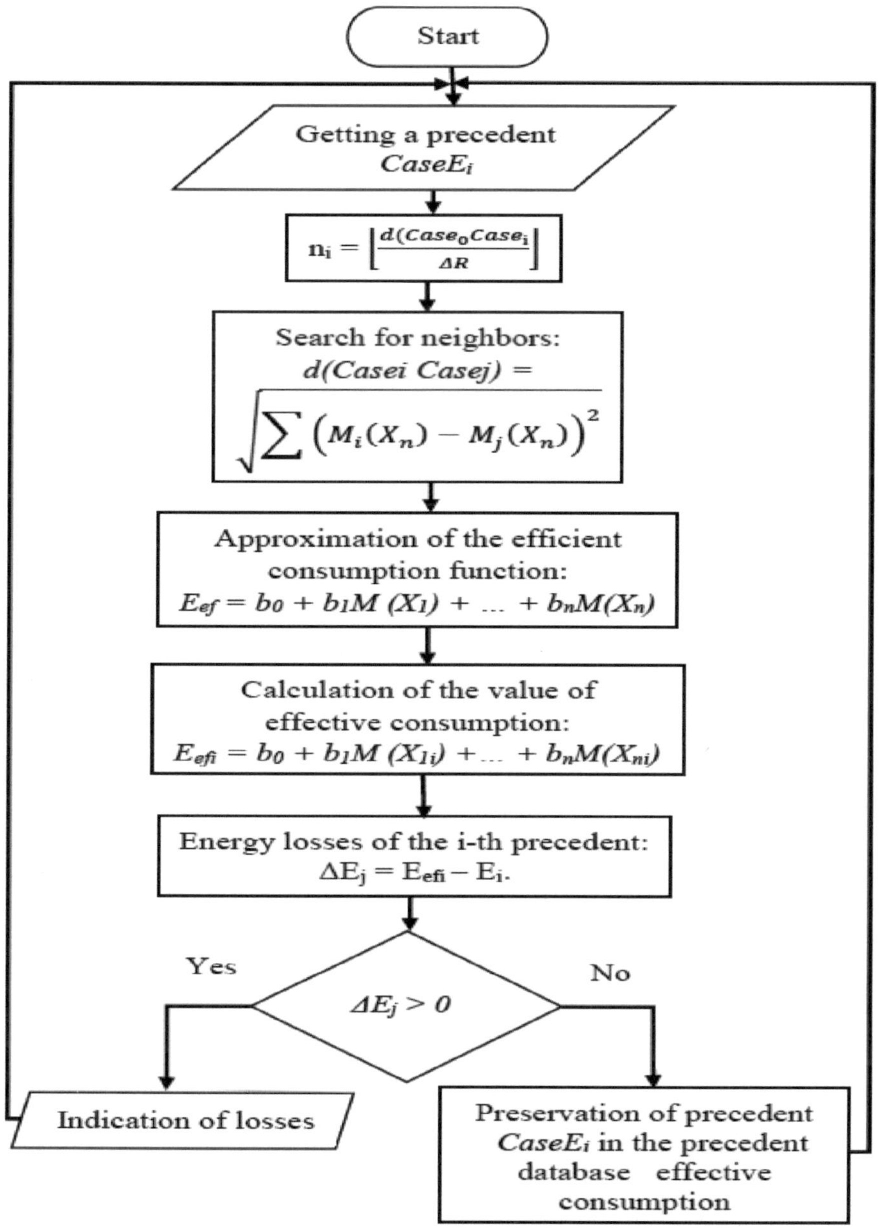

Figure 2.
Cyclic algorithm for forming a base of precedents and determining the amount of power of efficient energy consumption.

Produced ammonia, tone	Used natural gas, 1000 m^3/hour	Actually used electricity, 1000 kWh	Used electricity according to the regression model, 1000 kWh	Used electricity according to the precedent model, 1000 kWh
39,699	45,211	31,988	31,988	31,959
39,292	45,182	32,005	32,066	31,608
39,644	45,357	31,932	31,966	32,055
39,929	45,122	31,923	32,038	31,934
39,684	45,481	32,105	32,191	32,003
39,967	45,782	32,056	32,164	32,499
39,422	45,761	32,063	32,251	32,123
43,174	45,602	32,098	32,05	31,873
42,055	45,608	31,971	33,247	32,178
40,449	45,023	31,953	32,79	31,842
41,385	44,973	31,86	32,27	31,922
46,907	53,037	35,576	35,576	35,551
46,923	52,979	35,532	35,699	35,565
46,109	52,78	35,521	35,67	35,594
46,772	52,693	35,491	35,395	35,508
46,583	52,644	35,501	35,598	35,446
46,821	52,893	35,538	35,581	35,565
46,865	52,657	35,504	35,616	35,469

Table 1.
Comparison of regression and precedent approaches.

Numerical modeling of the method of precedent estimation of energy losses was performed on the data of chemical production given in [16]. The volumes of produced ammonia and consumed natural gas are accepted as factors influencing electricity consumption. The actual consumed electricity calculated according to the regression and precedent model is compared. The simulation results are shown in **Table 1**.

When building a precedent model, 18 precedents were considered. For each of the precedents, 4 nearest neighbors were selected, on which, by the method of least squares, the linear function of efficient energy consumption was built and the efficient energy consumption was calculated.

7. Conclusion

This chapter proposes a method for detecting and estimating latent and abnormal energy losses in production technology systems. The methodology is based on a retrospective analysis of cases of quasi-stationary energy consumption.

Author details

Borys Pleskach
Department of Energy Saving, Pukhov Institute for Modelling in Energy Engineering, Kyiv, Ukraine

*Address all correspondence to: bn.pleskach@gmail.com

References

[1] World Energy Outlook 2019, International Energy Agency, Paris. Available from: http://https://www.iea.org/reports/world-energy-outlook-2019

[2] Waide P., Brunner C.U. Energy-Efficiency Policy Opportunities for Electric Motor-Driven Systems: International Energy Agency, 2011.

[3] Soner Emec, Jörg Krüger, Günther Seliger. Online fault-monitoring in machine tools based on energy consumption analysis and non-invasive data acquisition for improved resource-efficiency. In: 13th Global Conference on Sustainable Manufacturing, Procedia CIRP 40, 2016, p. 236 – 243

[4] Bunse, K., Vodicka, M., Schönsleben, P., Brülhart, M., Ernst, F.O. Integrating energy efficiency performance in production management - gap analysis between industrial needs and scientific literature: Journal of Cleaner Production, 2011, 19, 667–679, https://doi.org/10.1016/j.jclepro.2010.11.011

[5] ShaohuaHuFeiLiu YanHeTongHu, An on-line approach for energy efficiency monitoring of machine tools: Journal of Cleaner Production, Volume 27, May 2012, pp. 133-140

[6] Berezhny C.V., O.Ye. Melnyk OE, Methods for determining the specific norms of electricity consumption: Engineering in agricultural production, industrial engineering, automation, 2012, issue 25, part II p. 145-155.

[7] Nakhodov VF, Borychenko OV, Ivanko DO Control of energy efficiency in the energy management system: Bulletin of KNUTD, 2013, №6, p. 67 - 76

[8] Efficiency Direct, Energy Monitoring and Targeting [Online]. Available from: https: //efficiency-direct.co.uk/services/energy-monitoring-and-targeting/

[9] Osipov G.S., Methods of artificial intelligence: M.: FIZMATLIT, 2011. 296 p.

[10] Eamonn Keogh, Selina Chu, David Hart, Michael Pazzani, Segmenting Time Series: A Survey and Novel Approach, 2003 Available from: https://www.cs.rutgers.edu/~pazzani/Publications/survey.pdf

[11] Shumway R.H., Stoffer D.S. Time Series Analysis and Its Applications: With R Examples, 3rd Edition. — Springer, 2011. — 609 p.

[12] Vasko K, Toivonen H, Estimating the number of segments in time series data using permutation tests. In: Proceedings of the IEEE International Conference on Data Mining, 2002, pp. 466–473

[13] Himberg J, Korpiaho K, Mannila H, Tikanmaki J, Toivonen H. Time-series segmentation for context recognition in mobile devices. In: Proceedings of the IEEE International Conference on Data Mining (ICDM01), 2001, San Jose, California pp. 466–473

[14] Kohonen T. Self-Organizing Maps: Springer, 1995. –501 p.

[15] Zmitrovich A.I., Intelligent information systems: Mn.: NTOOO "TetraSystems", 1997, 368 p.

[16] I.B. Stetsenko, Ya. S. Bederak, Construction of multifactor mathematical models of energy consumption in chemical production: Energy saving. Energy. Energy audit, 2013, № 7, C. 41 - 48.